AF531646

Food and Dietetics

Food and Dietetics

Shivani Rawat

RANDOM PUBLICATIONS
NEW DELHI (INDIA)

Food and Dietetics

ISBN 978-93-5111-525-0

Published in 2015 in India by

RANDOM PUBLICATIONS

4376-A/4B, Gali Murari Lal, Ansari Road
New Delhi-110 002
Phone : +9111-43580356, 011-23289044, 011-43142548
e-mail: sales@randompublications.com,
info@randompublications.com, randomexports@gmail.com

Reprinted 2021

Type Setting by : Friends Media, Delhi-110089
Digitally Printed at: Replika Press Pvt. Ltd.

Preface

Human nutrition is the provision to obtain the materials necessary to support life. In general, people can survive for two to eight weeks without food, depending on stored body fat and muscle mass. Survival without water is usually limited to three or four days. Lack of food remains a serious problem, with about 36 million people dying every year from causes directly or indirectly related to hunger. Childhood malnutrition is also common and contributes to the global burden of disease. However global food distribution is not equal, and obesity among some human populations has increased to almost epidemic proportions, leading to health complications and increased mortality in some developed, and a few developing countries. Obesity is caused by consuming more calories than are expended, with many attributing excessive weight gain to a combination of overeating of "unhealthy" (high fat, high sugar, high carbohydrate) foods and insufficient exercise.

Nutritional science investigates the metabolic and physiological responses of the body to diet. With advances in the fields of molecular biology, biochemistry, and genetics, the study of nutrition is increasingly concerned with metabolism and metabolic pathways: the sequences of biochemical steps through which substances in living things change from one form to another.

This book is written as a course text for those studying degree courses in nutrition and dietetics and for students on modular courses on nutrition within other degree courses, e.g. food studies, medicine, health sciences, nursing and biological sciences. It is also of great value as a reference for professional nutritionists and dietitians, food scientists and health professionals based in academia, in practice and in commercial positions such as within the food and pharmaceutical industries.

Author

Contents

1

Functions of Food

The phrase "we are what we eat" is frequently used to signify that the composition of our bodies is dependent in large measure on what we have consumed. The many chemical elements in the human body occur mainly in the form of water, protein, fats, mineral salts and carbohydrates, in the percentages shown in Table 1. Each human body is built up from food containing these five constituents, and vitamins as well.

Food serves mainly for growth, energy and body repair, maintenance and protection. Food also provides enjoyment and stimulation, since eating and drinking are among the pleasures of life everywhere. Truly food nourishes both the body and soul. Even if technology could produce a perfect diet in terms of content, such a diet could still lack, for example, the aroma and flavour of a curry, or the stimulating taste of hot coffee.

What controls appetite or the feeling of hunger is not fully understood. The hypothalamus in the brain has a role, as do other central nervous system sites. Other probable factors include blood sugar levels, body hormones, body fat, many diseases, emotions and of course food type and availability, personal likes and dislikes and the social setting where food is to be consumed.

Dietary Constituents

A simple classification of dietary constituents is given in Table 2. Human beings eat food, and not individual nutrients. Most foods, including staples such as rice, maize and wheat, provide mainly carbohydrate for energy but also significant quantities of protein, a little fat or oil and useful micronutrients. Thus cereal grains provide some of the constituents needed for energy, growth and body repair and maintenance. Breastmilk provides all the macro- and micronutrients necessary to satisfy the total needs of a young infant up to six months of age including those for energy, growth and body repair and maintenance. Cows' milk has the balance of nutrients for all the requirements of a calf.

Table 1. Chemical composition of a human body weighing 65 kg

Component	*Percentage of body weight*
Water	61.6
Protein	17
Fats	13.8
Minerals	6.1
Carbohydrate	1.5

Table 2.Simple classification of dietary constituents

Constituent	*Use*
Water	To provide body fluid and to help regulate body temperature
Carbohydrates	As fuel for energy for body heat and work
Fats	As fuel for energy and essential fatty acids
Proteins	For growth and repair
Minerals	For developing body tissues and for metabolic processes and protecfion
Vitamins	For metabolic processes and protection
Indigestible and unabsorbable particles, including fibre	To form a vehicle for other nutrients, add bulk to the diet, provide a habitat for bacterial flora and assist proper elimination of refuse

Water

Water can be considered the most important dietary constituent. A normal man or woman can live without food for 20 to 40 days, but without water humans die in four to seven days. Over 60 percent of human body weight is made up of water, of which approximately 61 percent is intracellular and the rest extracellular. Water intake, except under exceptional circumstances (e.g. intravenous feeding), comes from the food and fluids consumed. The amount consumed varies widely in individuals and may be influenced by climate, culture and other factors. Often as much as 1 litre is consumed in solid food, and 1 to 3 litres as fluids drunk. Water is also formed in the body as a result of oxidation of macronutrients, but the water thus obtained usually constitutes less than 10 percent of total water.

Water is excreted mainly by the kidneys as urine. The kidneys regulate the output of urine and maintain a balance; if smaller amounts of fluid are consumed, the kidneys excrete less water, and the urine is more concentrated. While most water is eliminated by the kidneys, in hot climates as much or more can be lost from the skin (through perspiration) and the lungs. Much smaller quantities are lost from the gut in the faeces (except in the presence of diarrhoea, when losses may be high).

Metabolism of sodium and potassium, which are known as electrolytes, is linked with body water. The sodium is mainly in the extracellular water and the potassium in the intracellular water. Most diets contain adequate amounts of both these minerals. In fluid loss caused, for example, by diarrhoea or haemorrhage, the balance of electrolytes in the blood may become disturbed. Water intake and electrolyte balance are particularly important in sick infants. In healthy infants, breastmilk alone from a healthy mother provides adequate quantities of fluids and electrolytes without additional water for the first six months of life even in hot climates. Infants with diarrhoea and disease, however, may require additional fluids.

While food intake is largely regulated by appetite and food availability, fluid intake is influenced by the sensation termed thirst. Thirst may arise for various reasons. In dehydration it may be caused by drying of the mouth but also by signals from the same satiety centre in the hypothalamus that controls hunger sensations. The phenomenon of water accumulation in the body is manifested in the condition known as oedema, when disease causes an excess of extracellular fluid. Two important deficiency diseases in which generalized oedema is a feature are kwashiorkor and wet beriberi. The excess fluid may result from electrolyte disturbances and accumulation of water in the extracellular compartment. A person can have oedema and still be dehydrated from diarrhoea; this condition is a form of heart failure. Water can also collect in the peritoneal cavity, in the condition known as ascites, which may be caused by liver disease.

Body Composition

The human body is sometimes said to be divided into three compartments, accounting for the following shares of the total body weight of a well-nourished healthy adult male:

— body cell mass, 55 percent;

— extracellular supporting tissue, 30 percent;

— body fat, 15 percent.

The body cell mass is made up of cellular components such as muscle, body organs (viscera, liver, brain, etc.) and blood. It comprises the parts of the body that are involved in body metabolism, body functioning, body work and so on.

The extracellular supporting tissue consists of two parts: the extracellular fluid (for example, the blood plasma supporting the blood cells) and the skeleton and other supporting structures.

Body fat is nearly all present beneath the skin (subcutaneous fat) and around body organs such as the intestine and heart. It serves in part as an energy reserve. Small quantities are present in the walls of body cells or in nerves.

Physiologists and those interested in metabolism have developed various ways to estimate body composition, including the amount of fluids in the body and body density. A common determination is to estimate lean body mass (LBM) or the fat-free mass of the body. These measures vary from the very simple to the very difficult. The simpler ones are of course less precise. Anthropometry using weight, height, skinfold thickness and body circumferences is relatively easy and very cheap to undertake, and does provide some estimate of LBM and body composition. In contrast, methods using, for example, bioelectrical impedance, computerized axial tomography (CAT scans) and nuclear magnetic resonance require expensive apparatus and highly trained staff.

The fluid in the cells (intracellular fluid) has mainly potassium ions, and the extracellular fluid is mainly a solution of sodium chloride. Both also have other ions. Total body water can be estimated using different methods including dilution techniques to measure, for example, plasma volume.

Body fat is estimated using different methods. Because a large portion of adipose tissue is present beneath the skin, it can be estimated by using a skinfold calliper to measure skinfold thickness in different sites. Another method is to weigh the person both in air and under water using a special apparatus and tank. This method really provides an estimate of body density.

The various methods of determining body composition are described in detail in textbooks of physiology or nutrition.

Body composition is much influenced by nutrition. The two extremes are the wasting of nutritional marasmus and starvation and the overweight of obesity. Body composition differs between the genders and, perhaps only slightly, among races. African Americans have been shown to have heavier skeletons than whites of the same body build in the United States. In females pregnancy and lactation influence body composition.

The body composition of children is influenced by their age and growth. Disturbances of growth resulting from nutritional deficiencies influence body composition, including the eventual size of the body and of body organs.

Metabolism and Energy

The general term for all the chemical processes carried out by the cells of the body is "metabolism". Chief among these processes is the oxidation (combustion, or burning) of food which produces energy. This process is analogous to a car engine burning petrol to produce the energy that makes it run. In most forms of combustion, be it in the car or in the human, heat is produced as well as energy.

Classical physics taught that energy can be neither created nor destroyed. Although this law of nature is not completely correct (as the conversion of matter to energy in a nuclear reactor shows), it is still true in most instances. All three macronutrients in food - carbohydrate, protein and fat provide energy. Energy for the body comes mainly from food, and in the absence of food it can be produced only by the breakdown of body tissues.

All forms of energy can be converted into heat energy. It is possible to measure the heat produced by burning a litre of petrol, for example. Food energy can also be and is expressed as heat energy. The unit of measurement used has been the large calorie (Cal) or kilocalorie (kcal) (which is 1 000 times the small calorie used in physics), but this measure is increasingly being replaced by the joule (J) or kilojoule (kJ). The kilocalorie is defined as the heat necessary to raise the temperature of 1 litre of water from 14.5° to 15.5°C. Whereas the kilocalorie is a unit of heat, the joule is truly a unit of energy. The joule is defined as the amount of energy used when 1 kg is moved 1 m by 1 newton (N) of force. In nutrition the kilojoule (1000 J) is used. The equivalent of 1 kcal is 4.184 kJ. These are units of measurement in the same way that litres and pints are measures of quantity, and metres and feet are measures of length. In many scientific journals the joule is being introduced in place of the kilocalorie, but the general public and most health workers still prefer to express food energy in kilocalories rather than joules. Kilocalories are therefore used.

The human body requires energy for all bodily functions, including work, the maintenance of body temperature and the continuous action of the heart and lungs In children energy is essential for growth Energy is also needed for breakdown, repair and building of tissues. These arc metabolic processes. The rate at which these functions are carried out while the body is at rest is the Basal metabolic rate (BMR).

Basal Metabolic Rate

BMR for an individual person is usually defined as the amount of energy [expressed in kilocalories or megajoules (MJ, per day] expended when the person is al complete rest, both physical (i.e. lying down) and psychological. It can also be expressed as kilocalories per hour or per kilogram of weight. BMR provides the energy required by the body for maintenance of body temperature; for the work of body organs such as the beating heart and the muscles working for normal, at rest, breathing; and for the functioning of other organs such as the liver, kidneys and brain.

BMR varies from individual to individual. Important general factors influencing BMR are the person's weight, gender, age and state of health. BMR is also influenced by the person's body composition, for example the amounts of muscle and adipose tissue and therefore the amounts of protein and fat in the body. In broad terms, bigger people with more muscle and larger body organs have higher BMR than smaller people. Elderly people tend to have lower BMR than they had when they were young, and females tend to have lower BMR than males even on a per kilogram body weight basis. There are exceptions, however, to all these generalizations.

BMR is important as a component of energy requirements. Table 3 shows BMR of adult men and women according to height and weight, both per kilogram body weight and as total energy per day. The table shows, for example, that in females aged 30 to 60 years BMR ranges from 1 190 to 1 420 kcal per day. This is the amount of energy required by a woman at

Table 3. Basal metabolic rate in adult men and women in relation to height and median acceptable

Height	*Weight*[a]	*18-30 years*		*30-60 years*		*Over 60*	
(m)	*(kg)*	*kcal(kJ)*[b]*/kg/day*	*kcal(kJ)/day*	*kcal(kJ)/kg/day*	*kcal(kJ)/day*	*kcal(kJ)/kg/day*	*kcal(kJ)/day*
Men							
1.5	49.5	29.0 (121)	1 440 (6.03)	29.4 (123)	1450 (6.07)	23.3 (98)	1150 (4.81)
1.6	56.5	27.4 (115)	1 540 (6.44)	27.2 (114)	1530 (6.40)	22.2 (93)	1250 (5.23)
1.7	63.5	26.0 (109)	1 650 (6.90)	25.4 (106)	1620 (6.78)	21.2 (89)	1 350 (5.65)
1.8	71.5	24.8 (104)	1 770 (7.41)	23.9 (99)	1710 (7.15)	20.3 (85)	1450 (6.07)
1.9	79.5	23.9 (100)	1 890 (7.91)	22.7 (95)	1 800 (7.53)	19.6 (82)	1560 (6.53)
2.0	88.0	23.0 (96)	2030 (8.49)	21.6 (90)	1 900 (7.95)	19.0 (80)	1 670 (6.99)
Women							
1.4	41	26.7 (112)	1100 (4.60)	28.8 (120)	1190 (4.98)	25.0 (105)	1 030 (4.31)
1.5	47	25.2 (105)	1 190 (4.98)	26.3 (110)	1 240 (5.19)	23.1 (97)	1090 (4.56)
1.6	54	23.9 (100)	1290 (5.40)	24.1 (101)	1 300 (5.44)	21.6 (90)	1160 (4.85)
1.7	61	22.9 (96)	1 390 (5.82)	22.4 (94)	1 360 (5.69)	20.3 (85)	1 230 (5.15)
1.8	68	22.0 (92)	1500 (6.28)	20.9 (87)	1 420 (5.94)	19.3 (81)	1 310 (5.48)

complete rest for 24 hours. Of course many adult females in developing countries are smaller than 1.4 m in height and 41 kg in weight; their BMR might then be a little lower than 1 190 kcal per day.

Energy Requirements

The mean daily energy requirements of adult men and women doing work classified as light, moderate and heavy are given in Table 4, expressed as multiples of BMR. The table shows, for example, that a woman doing heavy work requires energy equal to 1.32 times her BMR. If the woman is aged 25 years, is 1.4 m tall and weighs 41 kg, according to Table 3 her BMR would be 1100 kcal per day. Thus her daily requirements are: 1 100 kcal x 1.32 = 2 002 kcal.

Table 4. Average daily energy requirements of adults by category of occupational work expressed as a multiple of BMR

Classification of work	*Men*	*Women*
Light	1.55	1.56
Moderate	1.78	1.64
Heavy	2.10	1.82

It is often useful to estimate energy needs for various activities that a person may do for particular lengths of time. The energy expenditure is usually calculated by multiplying an activity factor or metabolic constant, which varies according to the activity, by the individual's BMR. Table 5 gives the activity factors for calculating gross energy expenditure of various activities for adult males and females.

Table 5.Activity factors for calculating gross energy expenditure (multiply by BMR)

Activity	*Adult males*	*Adult females*
Sleeping	1.0	1.0
Lying	1.2	1.2
Sitting quietly	1.2	1.2
Standing quietly	1.5	1.5
Walking slowly	2.8	2.8
Walking at normal pace	3.2	3.3
Walking fast uphill	7.5	6.6
Cooking	1.8	1.8

Office work (moving around)	1.6	1.7
Driving lorry	1.4	1.4
Labouring	5.2	4.4
Cutting sugar cane	6.5	-
Pulling loaded cart	5.9	-
Playing soccer	6.6	6.3
Fetching water from well	-	4.1
Pounding grain	-	4.6

The average human burns energy at his or her BMR only when at complete rest. All ordinary movements require additional energy, and physical work, of course, requires more still. For a healthy male with BMR of 1 kcal/min, an average day may involve the energy expenditure shown in Table 6.

Table 6. Energy expenditure of an average day for a healthy male

Activity	*Time (hours)*	*Energy expenditure (kcal/min)*	*Calculation*	*Total energy expenditure (kcal)*
Sleep	8	1 (=BMR)	88 x 60 x 1	480
Light work: herding animals	8	2.5	8 x 60 x 2.5	1 200
Other: sitting and minor activities	8	2	8 x 60 x 2	960
Total				*2 640*

If the person in this example did instead of eight hours of light work - five hours of herding and three hours of heavy work, hoeing hard ground at 8 kcal/min, then his output of energy would be as shown in Table 7.

If the individual undertaking the activities in the first example gets exactly 2 640 kcal in his food, his weight will be steady, and he will be functioning normally. However, if he then undertakes the activities in the second example and eats no extra food, his weight will gradually drop, because he will have to burn up his fuel reserve, which forms part of his own body. He would fairly soon, however, begin to limit his activities in order to stop this process. He would therefore probably work much less hard at hoeing, so that instead of burning 8 kcal/min he might use only, say, 3.2

kcal/min; he would also tend to be tired at the end of the day, and might well increase his period of complete rest (at 1 kcal/min) by reducing the period of minor activities. He would therefore have reduced his energy requirements to 2 646 kcal, as shown in Table 8.

Table 7. Energy expenditure when the person in Table 6 performs three hours of hard work

Activity	*Time (hours)*	*Energy expenditure (kcal/min)*	*Total energy expenditure (kcal)*
Sleep	8	1	480
Light work: herding	5	2.5	750
Hard work: hoeing	3	8	1440
Other: sitting and minor activities	8	2	960
Total			*3 630*

Table 8. Table Energy expenditure when the person in Table 7 adjusts his work to a less adequate diet

Activity	*Time (hours)*	*Energy expenditure (kcal/min)*	*Total energy expenditure (kcal)*
Sleep	10	1	600
Light work	5	2.5	750
Less hard work: hoeing	3	3.2	576
Other: sitting and minor activities	6	2	720
Total			*2 646*

This is just an example. In most in stances, when people increase their output of energy, including work, they feel more hungry and increase their consumption of their staple food, be it rice, millet, maize wheat, cassava or anything else.

The energy requirements of a human being are affected by several factors. The important ones are:

— *Body size. A* small person needs less energy than a large person.

— *Basal metabolic rate.* BMR varies and can be affected by factors such as disease of the thyroid gland.

— *Activity.* The more physical work or recreation performed, the more energy is required.

— *Pregnancy.* A woman requires extra energy to develop the foetus and to carry its additional weight.

— *Lactation.* The lactating mother needs additional energy to produce energy-containing milk for the suckling baby. The relatively long duration of breastfeeding among most Asians and Africans results in a large proportion of women requiring extra energy.

— *Age.* Infants and children need more energy, for growth and activity, than adults. In older persons, the need for energy is sometimes reduced because there is a decline in activity and because their BMR is usually lower.

— *Climate.* In warm climates, i.e. in most of the tropics and subtropics, less energy is necessary to keep the body at its normal temperature than in cold climates.

References

FAO.1981. *Traditional and non-traditionalfoods.* FAO Food and Nutrition Series No. 2. Rome.

King, F.S. & Burgess, A.1993. *Nutrition for developing countries.* Oxford, UK, Oxford University Press. 2nd ed.

Holland, B., Unwin, I.D. & Buss, D.H.1988. *Cereals and cereal products. Third supplement to McCance & Widdowson's The composition of foods.*Nottingham, UK, Royal Society of Chemistry.

McLaren, D.S. 1983. *Nutrition in the community.* New York, USA, John Wiley and Sons. 2nd ed.

2

Food Nutrients

\All living things need food to survive. It gives us energy for everything that we do. It also gives the body what it needs to repair muscles, organs and skin. Food helps us fight off dangerous diseases. It is important to eat a wide range of food in order to stay healthy. Nutrition is the science that deals with food and how the body uses it.

Food has nutrients in it— substances that give our body many important things that we need. They provide us with energy and also help control the way our body grows.

Before nutrients can go to work food must be broken down so that they can pass into our body. This is called digestion. It starts when we chew the food that we eat. When we swallow it it travels on to the stomach where it is mixed together with water and other fluids. Then the food is passed on to the intestine. Nutrients escape through the walls of the intestine into our blood. From there they are carried to all parts of the body.

Most food leaves waste that the body cannot use. Some of it goes to the kidneys and turns into urine. The liver also filters out waste. What is left over passes through the large intestine and leaves our body.

There are six main groups of nutrients: proteins, carbohydrates, fats, vitamins, minerals and water. The energy that food gives us is measured in kilocalories, or one thousand calories. A calorie is the energy that is needed to raise the temperature of water by one degree Celsius.

Carbohydrates

The main source of energy for most Asians, Africans and Latin Americans is carbohydrates in the food they eat. Carbohydrates constitute by far the greatest portion of their diet, as much as 80 percent in some cases. In contrast, carbohydrates make up only 45 to 50 percent of the diet of many people in industrialized countries.

Carbohydrates are compounds containing carbon, hydrogen and oxygen in the proportions 6:12:6. They are burned during metabolism to produce energy, liberating carbon dioxide (CO2) and water (H2O). The carbohydrates in the human diet are mainly in the form of starches and various sugars. Carbohydrates can be divided into three groups:

— monosaccharides, e.g. glucose, fructose, galactose;

— disaccharides, e.g. sucrose (table sugar), lactose, maltose;

— polysaccharides, e.g. starch, glycogen (animal starch), cellulose.

Monosaccharides

The simplest carbohydrates are the monosaccharides, or simple sugars. These sugars can pass through the wall of the alimentary tract without being changed by the digestive enzymes. The three most common are glucose, fructose and galactose.

Glucose, sometimes also called dextrose, is present in fruit, sweet potatoes, onions and other plant substances. It is the substance into which many other carbohydrates, such as the disaccharides and starches, are converted by the digestive enzymes. Glucose is oxidized to produce energy, heat and carbon dioxide, which is exhaled in breathing.

Because glucose is the sugar in blood, it is most often used as an energy-producing substance for persons fed intravenously. Glucose dissolved in sterile water, usually in concentrations of 5 or 10 percent, is frequently used for this purpose.

Fructose is present in honey and some fruit juices. Galactose is a monosaccharide that is formed, along with glucose, when the milk sugar lactose is broken down by the digestive enzymes.

Disaccharides

The disaccharides, composed of simple sugars, need to be converted by the body into monosaccharides before they can be absorbed from the alimentary tract. Examples of disaccharides are sucrose, lactose and maltose. Sucrose is the scientific name for table sugar (the kind that is used, for example, to sweeten tea). It is most commonly produced from sugar cane but is also produced from beets. Sucrose is also present in carrots and pineapple. Lactose is the disaccharide present in human and animal milk. It is much less sweet than sucrose. Maltose is found in germinating seeds.

Polysaccharides

The polysaccharides are chemically the most complicated carbohydrates. They tend to be insoluble in water, and only some can be used by human beings to produce energy. Examples of polysaccharides are starch, glycogen and cellulose.

Starch is an important source of energy for humans. It occurs in cereal grains as well as in root foods such as potatoes and cassava. Starch is liberated during cooking when the starch granules rupture because of heating.

Glycogen is made in the human body and is sometimes known as animal starch. It is formed from monosaccharides produced by the digestion of dietary starch. Starch from rice or cassava is broken down in the intestines to form monosaccharide molecules, which pass into the bloodstream. Those surplus monosaccharides that are not used to produce energy (and carbon dioxide and water) are fused together to form a new polysaccharide, glycogen. Glycogen is usually present in muscle and in the liver, but not in large amounts.

Any of the digestible carbohydrates when consumed in excess of body needs are converted by the body into fat which is laid down as adipose tissue beneath the skin and at other sites in the body.

Cellulose, hemicellulose, lignin, pectin and gums are sometimes called unavailable carbohydrates because humans cannot digest them. Cellulose and hemicellulose are plant polymers that are the main components of cell walls. They are fibrous substances. Cellulose, which is a polymer of glucose, is one of the fibres of green plants. Hemicellulose is a polymer of other sugars, usually hexose and pentose. Lignin is the main component of wood. Pectins are present in plant tissue and sap and are colloidal polysaccharides. Gums are also viscous carbohydrates extracted from plants. Pectins and gums are both used by the food industry. The human alimentary tract cannot break down these carbohydrates or utilize them to produce energy. Some animals, such as cattle, have microorganisms in their intestines that break down cellulose and make it available as an energy-producing food. In humans, any of the unavailable carbohydrates present in food pass through the intestinal tract. They form much of the bulk and roughage evacuated in human faeces, and are often termed "dietary fibre".

There is increasing interest in fibre in diets, because high-fibre diets are now considered healthful. A clear advantage of a high-fibre diet is a lower incidence of constipation than among people who consume a low-fibre diet. The bulk in high-fibre diets may contribute a feeling of fullness or satiety which may lead to less consumption of energy, and this may help reduce the likelihood of obesity. A high-fibre diet results in more rapid transit of food through the intestinal tract and is thus believed to assist normal and healthy intestinal and bowel functioning. Dietary fibre has also been found to bind bile in the intestines.

It is now recognized that the high fibre content of most traditional diets may be an important factor in the prevention of certain diseases which appear to be much more prevalent in people consuming the low-fibre diets common in industrialized countries. Because it facilitates the rapid passage of materials through the intestine, fibre may be a factor in the control of diverticulitis, appendicitis, haemorrhoids and also possibly arteriosclerosis, which leads to coronary heart disease and some cancers.

Frequent consumption of any sticky fermentable carbohydrates, either starch or sugar, can contribute to dental caries, particularly when coupled with poor oral hygiene. Adequate intake of fluoride and/or a topical application is the best protection against caries.

Fats

In many developing countries dietary fats make up a smaller part of total energy intake (often only 8 or 10 percent) than carbohydrates. In most industrialized countries the proportion of fat intake is much higher. In the United States, for example, an average of 36 percent of total energy is derived from fat.

Fats, like carbohydrates, contain carbon, hydrogen and oxygen. They are insoluble in water but soluble in such chemical solvents as ether, chloroform and benzene. The term "fat" is used here to include all fats and oils that are edible and occur in human diets, ranging from those that are solid at cool room temperatures, such as butter, to those that are liquid at similar temperatures, such as groundnut or cottonseed oils. (In some terminologies the word "oil" is used to refer to those materials that are liquid at room temperature, while those that are solid are called fats.)

Fats (also referred to as lipids) in the body are divided into two groups: storage fat and structural fat. Storage fat provides a reserve storehouse of fuel for the body, while the structural fats are part of the essential structure of the cells, occurring in cell membranes, mitochondria and intracellular organelles.

Cholesterol is a lipid present in all cell membranes. It has an important role in fat transport and is the precursor from which bile salts and adrenal and sex hormones are made.

Dietary fats consist mainly of triglycerides, which can be split into glycerol and chains of carbon, hydrogen and oxygen called fatty acids. This action, the digestion or breakdown of fats, is achieved in the human intestine by enzymes known as lipases, which are present primarily in the pancreatic and intestinal secretions. Bile salts from the liver emulsify the fatty acids to make them more soluble in water and hence more easily absorbed.

The many fatty acids in human diets are divided into two main groups: saturated and unsaturated. The latter group includes both polyunsaturated and mono-unsaturated fatty acids. Saturated fatty acids have the maximum number of hydrogen atoms that their chemical structure will permit. All fats and oils eaten by humans are mixtures of saturated and unsaturated fatty acids. Broadly speaking, fats from land animals (i.e. meat fat, butter and ghee) contain more saturated fatty acids than do those of vegetable origin. Fats from plant products and to some extent those from fish have more unsaturated fatty acids, particularly polyunsaturated fatty acids (PUFAs). There are exceptions, however. For example, coconut oil has a large amount of saturated fatty acids.

These groupings of fats have important health implications because excess intake of saturated fats is one of the risk factors associated with arteriosclerosis and coronary heart disease. In contrast, PUFAs are believed to be protective.

PUFAs also include two unsaturated fatty acids, linoleic acid and linolenic acid, which have been termed "essential fatty acids" (EFAs) as they are necessary for good health. EFAs are important in the synthesis of many cell structures and several biologically important compounds. Recent studies have also shown the benefits of other longer-chain fatty acids in the growth and development of young children, and arachidonic acid and docosa-hexaenoic acid (DHA) should conditionally be considered essential during early development. Experiments with animals and studies in humans have shown definite skin and growth changes and abnormal vascular and neural function in the absence of these fatty acids, and there is no doubt that they are essential for the nutrition of individual cells and tissues of the body.

Fat is desirable to make the diet more palatable. It also yields about 9 kcal/g, which is more than twice the energy yielded by carbohydrates and proteins (about 4 kcal/g); fat can therefore reduce the bulk of the diet. A person doing very heavy work, especially in a cold climate, may require as many as 4 000 kcal a day. In such a case it is highly desirable that a good proportion of the energy should come from fat; otherwise the diet would be very bulky. Bulky diets can be a particularly serious problem in young children as well. A reasonable increase in the fat or oil content of the diets of young children raises the energy density of predominantly bulky carbohydrate diets and is highly desirable.

Fat also functions as a vehicle that assists the absorption of fat-soluble vitamins.

Thus fats, and even specific types of fat, are essential to health. However, practically all diets provide the small amount required.

Fat deposited in the human body serves as a reserve fuel. It is an economic way of storing energy, because, as mentioned above, fat yields about twice as much energy, weight for weight, as does carbohydrate or protein. Fat is present beneath the skin as an insulation against cold, and it forms a supporting tissue for many organs such as the heart and intestines.

All fat in the body is not necessarily derived from fat that has been eaten. However, excess calories from the carbohydrate and protein in, for example, maize, cassava, rice or wheat can be converted into fat in the human body.

Proteins

Like carbohydrates and fats, proteins contain carbon, hydrogen and oxygen, but they also contain nitrogen and often sulphur. They are particularly important as nitrogenous substances, and are necessary for growth and repair of the body. Proteins are the main structural constituents of the cells and tissues of the body, and they make up the greater portion of the substance of the muscles and organs (apart from water). The proteins in different body tissues are not all exactly the same. The proteins in liver, in blood and in specific hormones, for example, are all different.

Proteins are necessary:

— for growth and development of the body;

— for body maintenance and the repair and replacement of worn out or damaged tissues;

— to produce metabolic and digestive enzymes;

— as an essential constituent of certain hormones, such as thyroxine and insulin.

Although proteins can yield energy, their main importance is rather as an essential constituent of all cells. All cells may need replacement from time to time, and their replacement requires protein.

Any protein eaten in excess of the amount needed for growth, cell and fluid replacement and various other metabolic functions is used to provide energy, which the body obtains by changing the protein into carbohydrate. If the carbohydrate and fat in the diet do not provide adequate energy, then protein is used to provide energy; as a result less protein is available for growth, cell replacement and other metabolic needs. This point is especially important for children, who need extra protein for growth. If they get too little food for their energy requirements, then the protein will be diverted for daily energy needs and will not be used for growth.

Amino Acids

All proteins consist of large molecules which are made of amino acids. The amino acids in any protein are linked together in chains, called peptide linkages. The various proteins are made of different amino acids linked together in different chains. Because there are many different amino acids, there are many different possible configurations, so there are many different proteins.

During digestion proteins break down to form amino acids much as complex carbohydrates such as starches break down into simple monosaccharides and fats break down into fatty acids. In the stomach and intestines various proteolytic enzymes hydrolyse the protein, releasing amino acids and peptides.

Plants are able to synthesize amino acids from simple inorganic chemical substances. Animals do not have this ability; they derive all the amino acids necessary for building their protein from consumption of plants or animals. As the animals eaten by humans initially derived their protein from plants, all amino acids in human diets have originated from this source.

Animals have differing abilities to convert one amino acid into another. In the human this ability is limited. Conversion occurs mainly in the liver. If the ability to convert one amino acid into another were unlimited, then the question of the protein content of diets and the prevention of protein deficiency would be simple. It would be enough merely to supply sufficient protein, irrespective of the quality or amino acid content of the protein supplied.

Of the large number of amino acids, 20 are common in plants and animals. Of these, eight have been found to be essential for the adult human and have thus been termed "essential amino acids" or "indispensable amino acids", namely: phenyl-alanine, tryptophan, methionine, lysine, leucine, isoleucine, valine and threonine. A ninth amino acid, histidine, is required for growth and is essential for infants and children; it may also be necessary for tissue repair. Other amino acids include glycine, alanine, serine, cystine, tyrosine, aspartic acid, glutamic acid, proline, hydroxyproline, citrulline and arginine. Each protein in a food is composed of a particular mixture of amino acids which might or might not contain all eight of the essential ones.

Protein Quality and Quantity

To assess the protein value of any food it is useful to know how much total protein it contains, which amino acids it has and how many essential amino acids are present and in what proportion. Much is now known about the individual proteins present in various foods, their amino acid content and therefore their quality and quantity. Some have a better mixture of amino acids than others, and these are said to have a higher biological value. The proteins albumin in egg and casein in milk, for example, contain all the essential amino acids in good proportions and are nutritionally superior to such proteins as zein in maize, which contains little tryptophan or lysine, and the protein in wheat, which contains only small quantities of lysine. It is not true, however, to say that the proteins in maize and wheat are not valuable. Although they contain less of certain amino acids, they do contain some amount of all the essential amino acids as well as many of the other important ones. The relative deficiency of maize and wheat proteins can be overcome by providing other foodstuffs containing more of the limited amino acids. It is therefore possible for two foods with low-value protein to complement each other to form a good protein mixture when eaten together.

Humans, especially children on diets deficient in animal protein, require a variety of foods of vegetable origin, not just one staple food. In many diets, pulses or legumes such as groundnuts, beans and cowpeas, though short of sulphur-containing amino acids, supplement the cereal proteins, which are often short of lysine. A mixture of foods of vegetable origin, especially if taken at the same meal, can serve as a substitute for animal protein.

FAO has produced tables showing the content of essential amino acids in different foodstuffs, from which it can be seen which foods best complement each other. It is also necessary, of course, to ascertain the total quantity of protein and amino acids in any food.

The quality of the protein depends largely on its amino acid composition and its digestibility. If a protein is deficient in one or more essential amino acids, its quality is lower. The most deficient of the essential amino acids in a protein is called the "limiting amino acid". The limiting amino acid determines the efficiency of utilization of the protein present in a food or combination of foods. Human beings usually eat food in meals which contain many proteins; they seldom consume just one protein. Therefore nutritionists are interested in the protein quality of a person's diet or meals, rather than just one food. If one essential amino acid is in short supply in the diet, it limits the use of the other amino acids for building protein.

Readers who wish to become familiar with the methods used for determining protein quality are advised to consult comprehensive textbooks on nutrition, which describe them in detail. One method uses experiments on growth and nitrogen retention in young rats. Another involves determination of the amino acid or chemical score, usually by examining the efficiency of utilization of proteins in the foods consumed by comparing their amino acid composition with that of protein known to be of high quality, such as that in whole eggs.

The chemical score may thus be defined as the efficiency of utilization of food protein in comparison with whole egg protein. Net protein utilization (NPU) is a measure of the amount or percentage of protein retained in relation to that consumed. As an example, Table 1 gives the chemical score and NPU of the protein in five foods.It is not usual or easy to obtain NPU values in people, and in most studies rats are used.

Tabel 1. Chemical score and net protein utilization in selected foods

Food	*Chemical score*	*NPU determined in children*	*NPU determined in rats*
Eggs (whole)	100	87	94
Milk (human)	100	94	87
Rice	67	63	59
Maize	49	36	52
Wheat	53	49	48

Table 1 suggests that there is a good correlation between the values in rats and in children, and that chemical score provides a reasonable estimate of protein quality.

For the professional involved in nutritional activities to help people - be it a dietitian in a health facility, an agricultural extension worker or a nutrition educator what is important is that the protein value differs among foods and that mixing foods improves the protein quality of the meal or the diet.

Table 2 gives the protein content and the limiting amino acid score of some commonly eaten plant-based foods. Because Iysine is most commonly the limiting amino acid in many foods of plant origin, the Iysine score is also given.

Protein Digestion and Absorption

Proteins consumed in the diet undergo a series of chemical changes in the gastrointestinal tract. The physiology of protein digestion is complicated; pepsin and rennin from the stomach, trypsin from the pancreas and erepsin from the intestines hydrolyse proteins into their component amino acids. Most of the amino acids are absorbed into the bloodstream from the small intestine and thus travel to the liver and from there all over the body. Any surplus amino acids are stripped of the amino (NH_2) group, which goes to form urea in the urine, leaving the rest of the molecule to be transformed into glucose. There is now some evidence that a little intact protein is taken up into certain cells lining the intestines. Some of this protein in the infant may have a role in the passive immunity conveyed from the mother to her newborn child.

A little of the protein and amino acids released in the intestines is not absorbed. The unabsorbed amino acids, plus cells shed from the intestinal villi and acted upon by bacteria, together with gut organisms, contribute to the nitrogen found in faeces.

Much of the protein in the human body is present in muscle. There is no true storage of protein in the body as there is with fat and to a small extent glycogen. However, there is now little doubt that a well-nourished individual has sufficient protein accumulated to be able to last several days without replenishment and to remain still in good health.

Table 2. Protein content, limiting amino acid score and Iysine score of selected plant foods

Food	*Protein content (%)*	*Limiting amino add score*	*Lysine score*
Cereals			
Maize	9.4	49 (Lys)	49
Rice (white)	7.1	62 (Lys)	62
Wheat flour	10.3	38 (Lys)	38
Millet	11.0	33 (Lys)	33
Legumes			
Kidney beans	23.6	100	118
Cowpea	23.5	100	117
Groundnut	25.8	62 (Lys)	62
Vegetables			
Tomato	0.9	56 (Leu)	64
Squash	1.2	70 (Thr)	95
Pepper,sweet	0.9	77 (Lys Leu)	77
Cassava	1.3	44 (Leu)	56
Potato	2.1	91 (Leu)	105

Protein Requirements

Children need more protein than adults because they need to grow. Infants in the first few months of life require about 2.5 g of protein per kilogram of body weight. This requirement drops to about 1.5 g/kg at nine to 12 months of age. Unless energy intakes are adequate, however, the protein will not all be used for growth. A pregnant woman needs an additional supply of protein to build up the foetus inside her. Similarly, a lactating woman needs extra protein, because the milk she secretes contains protein. In some societies it is common for women to breastfeed their babies for as long as two years. Thus some women need extra protein for two years and nine months for every infant they bear.

Protein requirements and recommended allowances have been the subject of much research, debate and disagreement over the past 50 years. FAO and the World Health Organization (WHO) periodically assemble experts to review current knowledge and to provide guidelines.

The most recent guidelines were the outcome of an Expert Consultation held jointly by FAO, WHO and United Nations University (UNU) in Rome in 1981 (WHO, 1985). The safe level of intake for a one-year-old child was put at 1.5 g per kilogram of body weight. The amount then falls to 1 g/kg at age six years. The United States recommended dietary allowance (RDA) is a little higher, namely 1.75 g/kg at age one year and 1.2 g/kg at age six years. In adults the FAO/WHO/UNU safe intake of protein is 0.8 g/kg for females and 0.85 g/kg for males.

Values are provided both for a diet high in fibre, comprising mainly cereals, roots and legumes with little food of animal origin, and for a mixed balanced diet with less fibre and plenty of complete protein. As an example, a non-pregnant adult woman weighing 55 kg requires 49 g of protein per day for the first diet and 41 g per day for the second. Fibre reduces protein utilization.

Inadequate protein intake jeopardizes growth and repair in the body. Protein deficiency is especially dangerous for children because they are growing and also because the risk of infection is greater during childhood than at almost any other time of life. In children inadequate energy intake also has an impact on protein. As stated above, in the absence of adequate energy some protein needs to be diverted and therefore will not be used for growth.

In many developing countries (though not all), the intake of protein is relatively low and of predominantly vegetable origin. The paucity of foods of animal origin in the diet is not always a matter of choice. For example, many low-income Africans and Latin Americans like animal products but find them less freely available, more difficult to produce and store and more expensive than most vegetable products. Diets low in meat, fish and dairy products are very common in countries where most people are poor.

Infections lead to an increased loss of nitrogen from the body, which has to be replaced by proteins in the diet. Therefore children and others who have frequent infections will have greater protein needs than healthy persons. This fact must constantly be borne in mind, for in developing countries many children suffer an almost continual series of infectious diseases; they may frequently get diarrhoea, and they may harbour intestinal parasites.

Minerals

Minerals have a number of functions in the body. Sodium, potassium and chlorine are present as salts in body fluids, where they have a physiological role in maintaining osmotic pressure. Minerals form part of the constitution of many tissues. For example, calcium and phosphorus in bones combine to give rigidity to the whole body. Minerals are present in body acids and alkalis; for example, chlorine occurs in hydrochloric acid in the stomach. They are also essential constituents of certain hormones, e.g. iodine in the thyroxine produced by the thyroid gland.

The principal minerals in the human body are calcium, phosphorus, potassium, sodium, chlorine, sulphur, copper, magnesium, manganese, iron, iodine, fluorine, zinc, cobalt and selenium. Phosphorus is so widely available in plants that a shortage of this element is unlikely in any diet. Potassium, sodium and chlorine are easily absorbed and are physiologically more important than phosphorus. Sulphur is consumed by humans mainly in the form of sulphur-containing amino acids; thus sulphur deficiency, when it occurs, is linked with protein deficiency. Copper, manganese and magnesium deficiencies are not believed to be common. The minerals that are of most importance in human nutrition are thus calcium, iron, iodine, fluorine and zinc, and only these are discussed in some detail here. Some mineral elements are required in very tiny amounts in human diets but are still vital for metabolic purposes; these are termed "essential trace elements".

Calcium

The body of an average-sized adult contains about 1 250 g of calcium. Over 99 percent of the calcium is in the bones and teeth, where it is combined with phosphorus as calcium phosphate, a hard substance that gives the body rigidity. However, although hard and rigid, the skeleton of the body is not the unchanging structure it appears to be. In fact, the bones are a cellular matrix, and the calcium is continuously taken up by the bones and given back to the body. The bones, therefore, serve as a reserve supply of this mineral.

Calcium is present in the serum of the blood in small but important quantities, usually about 10 mg per 100 ml of serum. There are also about 10 g of calcium in the extracellular fluids and soft tissues of the adult body.

Properties and functions

In humans and other mammals, calcium and phosphorus together have an important role as major components of the skeleton. They are also important, however, in metabolic functions such as muscular function, nervous stimuli, enzymatic and hormonal activities and transport of oxygen. These functions are described in detail in textbooks of physiology and nutrition.

The skeleton of a living person is physiologically different from the dry skeleton in a grave or museum. The bones are living tissues, consisting mainly of a mineralized protein collagen substance. In the living body there is continuous turnover of calcium. Bone is laid down and resorbed all the time, in people of all ages. Bone cells called osteoclasts take up or resorb bone, while others, termed osteoblasts, lay down or form new bone. The bone cells in the mineralized collagen are called osteocytes.

Up to full growth or maturity (which has usually taken place by age 18 to 22 years), new bone is formed as the skeleton enlarges to its adult size. In young adults, despite bone remodelling, the skeleton generally maintains its size. However, as persons get older there is some loss of bone mass.

A complex physiological system maintains proper calcium and phosphorus levels. The control involves hormones from the parathyroid gland, calcitonin and the active form of vitamin D (1,25-dihydroxy-cholecalciferol).

Small but highly important amounts of calcium are present in extracellular fluids, particularly blood plasma, as well as in various body cells. In serum most of the calcium is in two forms, ionized and protein bound. Laboratories usually measure only total plasma calcium; the normal range is 8.5 to 10.5 mg/dl (2.1 to 2.6 mmol/litre). A drop in the level of calcium to below 2.1 mmol/litre is termed hypocalcaemia and can lead to various symptoms. Tetany (not to be confused with tetanus resulting from the tetanus bacillus), characterized by spasms and sometimes fits, results from low levels of ionized calcium in the blood.

Dietary sources

All the calcium in the body, except that inherited from the mother, comes from food and water consumed. It is especially necessary to have adequate quantities of calcium during growth, for it is at this stage that the bones develop.

The foetus in the mother's uterus has most of its nutritional requirements satisfied, for in terms of nutrition the unborn child is almost parasitic. If the mother's diet is poor in calcium, she draws extra supplies of this mineral from her bones.

An entirely breastfed infant will obtain adequate calcium from breastmilk as long as the volume of milk is sufficient. Contrary to popular belief, the calcium content of human milk varies rather little; 100 ml of breastmilk, even from an undernourished mother on a diet very low in calcium, provides approximately 30 mg of calcium (Table 3). A lactating mother secreting 1 litre will thus lose 300 mg of calcium per day.

Table 3. Calcium content of various milks commonly used in developing countries

Source of milk	*Calcium content (mg/100 ml)*
Human	32
Cow	119
Camel	120
Goat	134
Water buffalo	169
Sheep	193

Cows' milk is a very rich source of calcium, richer than human milk. Whereas a litre of human milk contains 300 mg of calcium, a litre of cows' milk contains 1200 ma. The difference arises because a cow has to provide for her calf, which grows much more rapidly than a human infant and needs extra calcium for the hardening of its fast-growing skeleton. Similarly, the milk of most other domestic animals has a higher calcium content than human milk. This does not mean, however, that a child would be better off drinking cows' milk rather than human milk. Cows' milk yields more calcium than a child needs. A child (or even a baby) who drinks large quantities of cows' milk excretes any excess calcium, so it is of no benefit; it does not increase the child's growth rate beyond what is optimal.

Milk products such as cheese and yoghurt are also rich sources of calcium. Small saltwater and freshwater fish such as sardines and sprats supply good quantities of calcium since they are usually eaten whole, bones and all. Small dried fish known as dagaa in the United Republic of Tanzania, kapenta in Zambia and chela in India add useful calcium to the diet. Vegetables and pulses provide some calcium. Although cereals and roots are relatively poor sources of calcium, they often supply the major portion of the mineral in tropical diets by virtue of the quantities consumed.

The calcium content of drinking-water varies from place to place. Hard water usually contains high levels of calcium.

Absorption and utilization

The absorption of calcium is variable and generally rather low. It is related to the absorption of phosphorus and the other important mineral constituents of the bones. Vitamin D is essential for the proper absorption of calcium. Thus a person seriously deficient in vitamin D absorbs too little calcium, even if the intake of calcium is more than adequate, and could have a negative calcium balance. Phytates, phosphates and oxalates in food reduce calcium absorption.

Persons customarily consuming diets low in calcium appear to have better absorption of calcium than those on high-calcium diets. Unabsorbed calcium is excreted in the faeces. Excess calcium is excreted in the urine and in sweat.

Requirements

It is not easy to state categorically the human requirements for calcium, because there are several factors influencing absorption and considerable variations in calcium losses among individuals.

Needs for calcium are increased during pregnancy and lactation, and children require more calcium because of growth. Those on high-protein diets require more calcium in the diet.

The following are recommended levels of daily calcium intake:

— adults, 400 to 500 mg;
— children, 400 to 700 mg;
— pregnant and lactating women, 800 to 1 000 mg.

Deficiency states

Disease or malformation caused primarily by dietary deficiency of calcium is rare. There is little convincing evidence to show that the many diets of adults in developing countries supplying perhaps only 250 to 300 mg of calcium daily are harmful to health. It is assumed that adults achieve some sort of balance when intakes of calcium are low. Females who go through a series of pregnancies and long lactations may lose calcium and be at risk of osteomalacia. However, vitamin D deficiency, not calcium deficiency, is more often implicated in this condition.

In children the development of rickets results from vitamin D deficiency, not from dietary lack of calcium, in spite of increased calcium requirements in childhood. Calcium balance in childhood is generally positive, and calcium deficiency has not been shown to have an adverse influence on growth.

Osteoporosis is a common disease of ageing, especially in women. The skeleton becomes demineralized, which leads to fragility of bones and commonly to fractures of the hip, vertebrae and other bones, particularly in older women. High calcium intake is often recommended but has not been proved effective in prevention or treatment.

Exercise appears to reduce the loss of calcium from bones; this may explain, in part, why osteoporosis is less prevalent in many developing countries, where women work hard and are very active. There is now clear evidence that providing the female hormone oestrogen to women after menopause reduces bone loss and osteoporosis.

Iron

Iron deficiency is a very common cause of ill health in all parts of the world, both South and North. The average iron content in a healthy adult is only about 3 to 4 g, yet this relatively small quantity is vital.

Properties and functions

Most of the iron in the body is present in the red blood cells, mainly as a component of haemoglobin. Much of the rest is present in myoglobin, a compound occurring mainly in muscles, and as storage iron or ferritin, mainly in the liver, spleen and bone marrow. Additional tiny quantities are found binding protein in the blood plasma and in respiratory enzymes.

The main, vital function of iron is in the transfer of oxygen at various sites in the body. Haemoglobin is the pigment in the erythrocytes that carries oxygen from the lungs to the tissues. Myoglobin in skeletal and heart muscle accepts the oxygen from the haemoglobin. Iron is also present in peroxidase, catalase and the cytochromes.

Iron is an element that is neither used up nor destroyed in the properly functioning body. Unlike some minerals, it is not required for excretion, and only very small amounts appear in urine and sweat. Minute quantities are lost in desquamated cells from the skin and intestine, in shed hair and nails and in the bile and other body secretions. The body is, however, efficient, economical and conservative in the use of iron. Iron released when the erythrocytes are old and broken down is

taken up and used again and again for the manufacture of new erythrocytes. This economy of iron is important. In normal circumstances, only about 1 mg of iron is lost from the body daily by excretion into the intestines, in urine, in sweat or through loss of hair or surface epithelial cells.

Because iron is conserved, the nutritional needs of healthy males and postmenopausal females are very small. Women of child-bearing age, however, must replace the iron lost during menstruation and childbirth and must meet the additional requirements of pregnancy and lactation. Children have relatively high needs because of their rapid growth, which involves increases not only in body size but also in blood volume.

Dietary sources

Iron is present in a variety of foods of both plant and animal origin. Rich food sources include meat (especially liver), fish, eggs, legumes (including a variety of beans, peas and other pulses) and green leafy vegetables. Cereal grains such as maize, rice and wheat contain moderate amounts of iron, but because these are often staple foods and eaten in large quantities, they provide most of the iron for many people in developing countries. Iron cooking pots may be a source of iron.

Milk, contrary to the notion that it is the "perfect food", is a poor source of iron. Human milk contains about 2 mg of iron per litre and cows' milk only half this amount.

Absorption and utilization

Absorption of iron takes place mainly in the upper portion of the small intestine. Most of the iron enters the bloodstream directly and not through the lymphatic system. Evidence indicates that absorption is regulated to some extent by physiological demand. Persons who are iron deficient tend to absorb iron more efficiently and in greater quantities than do normal subjects.

Several other factors affect iron absorption. For example, tannins, phosphates and phytates in food reduce iron absorption, whereas ascorbic acid increases it. Studies have indicated that egg yolk, despite its relatively high iron content, inhibits absorption of iron - not only the iron from the egg yolk itself, but also that from other foods.

Healthy subjects normally absorb only 5 to 10 percent of the iron in their foods, whereas iron-deficient subjects may absorb twice that amount. Therefore, on a diet that supplies 15 mg of iron, the normal person would absorb 0.75 to 1.5 mg of iron, but the iron-deficient person would absorb as much as 3 mg. Iron absorption generally increases during growth and pregnancy, after bleeding and in other conditions in which the demand for iron is enhanced.

Of greatest importance is the fact that the availability of iron from foods varies widely. Absorption of the haem iron in foods of animal origin (meat, fish and poultry) is usually very high, whereas the non-haem iron in foods such as cereals, vegetables, roots and fruits is poorly absorbed.

However, people usually eat meals, not single individual foods, and a small amount of haem iron consumed with a meal where most of the iron is non-haem iron will enhance the absorption of all the iron. Thus the addition of a quite small amount of haem iron from perhaps fish or meat to a large helping of rice or maize containing non-haem iron will result in much greater absorption of iron from the cereal staple. If this meal also includes fruits or vegetables, the vitamin C in them will also enhance iron absorption. However, if tea is consumed with this meal, the tannin present in the tea will reduce the absorption of iron.

Requirements

The dietary requirements for iron are approximately ten times the body's physiological requirements. If a normally healthy man or post-menopausal woman requires 1 mg of iron daily because of iron losses, then the dietary requirements are about 10 mg per day. This recommendation allows a fair margin of safety, as absorption is increased with need.

Menstrual loss of iron has been estimated to average a little less than 1 mg per day during an entire year. It is recommended that women of child-bearing age have a dietary intake of 18 mg per day.

During pregnancy, the body requires on average about 1.5 mg of iron daily to develop the foetus and supportive tissues and to expand the maternal blood supply. Most of this additional iron is required in the second and third trimesters of pregnancy.

Breastfeeding women use iron to provide the approximately 2 mg of iron per litre of breastmilk. However, during the first six to 15 months of intensive breastfeeding they may not menstruate, so they do not lose iron in menstrual blood.

Newborn infants are born with very high haemoglobin levels (a high red blood cell count), termed polycythaemia, which provides an extra store of iron. This iron, together with that present in breastmilk, is usually sufficient for the first four to six months of life, after which iron from other foods becomes necessary.

Premature and other low-birth-weight infants may have lower iron stores and be at greater risk than other infants.

An excess intake of iron over long periods can lead to the disease siderosis or haemachromatosis. This disease is reported to occur most commonly where beer or other alcoholic beverages are brewed in iron cooking pots, particularly in South Africa. In alcoholics siderosis leading to iron deposits in the liver may be associated with cirrhosis.

Deficiency states

Consideration of the iron requirements and the iron content of commonly eaten foods might suggest that iron deficiency is rare, but this is not the case. Food iron is poorly absorbed. Iron is not readily excreted into the urine or the gastro-intestinal tract; thus severe iron deficiency is usually associated with an increased need for iron resulting from conditions such as pregnancy, blood loss or expansion of the total body mass during growth. Iron deficiency is most common in young children, in women of child-bearing age and in persons with chronic blood loss.

Hookworm infections, which are extremely prevalent in many countries, result in loss of blood which may cause iron deficiency anaemia. In some parts of the tropics schistosomiasis is also common, and this disease also causes blood loss.

Iodine

The body of an average adult contains about 20 to 50 mg of iodine, much of it in the thyroid gland. Iodine is essential for the formation of thyroid hormones secreted by this gland.

Properties and functions

In humans iodine functions as an essential component of the hormones of the thyroid gland, an endocrine gland situated in the lower neck. Thyroid hormones, of which the most important is thyroxine (T4), are important for regulating metabolism. In children they support normal growth and development, including mental development.

Iodine is absorbed from the gut as iodide, and excess is excreted in the urine. The adult thyroid gland, in a person consuming adequate iodine, traps about 60 μg of iodine per day to make normal amounts of thyroid hormones. If there is insufficient iodine, the thyroid works harder to trap more; the gland enlarges in size (a condition known as goitre), and its iodine content might become markedly reduced.

Thyroid stimulating hormone (TSH) from the pituitary gland influences thyroxine secretion and iodine trapping. In severe iodine deficiency, TSH levels are raised and thyroxine levels are low.

Dietary sources

Iodine is widely present in rocks and soils. The quantity in different plants varies according to the soil in which they are grown. It is not meaningful to list the iodine content of foodstuffs because of the large variations in iodine content from place to place, depending on the iodine content of the soil. Iodine tends to get washed out of the soil, and throughout the ages a considerable quantity has flowed into the sea. Sea fish, seaweed and most vegetables grown near the sea are useful sources of iodine. Drinking-water provides some iodine but very seldom enough to satisfy human requirements.

In many countries where goitre is prevalent the authorities have added iodine to salt, a strategy which has successfully controlled iodine deficiency disorders (IDD). Iodine has usually been added to salt in the form of potassium iodide, but another form, potassium iodate, is more stable and is better in hot, humid climates. Iodated salt is an important dietary source of iodine.

Deficiency states

A lack of iodine in the diet results in several health problems, one of which is goitre, or enlargement of the thyroid gland. Goitre is extremely prevalent in many countries. There are other contributing causes of goitre, but iodine deficiency is by far the most common. Iodine deficiency during pregnancy may lead to cretinism, mental retardation and other problems, which may be permanent, in the child. It is now known that endemic goitre and cretinism are not the only problems caused by iodine deficiency. The decrease in mental capacity associated with iodine deficiency is of particular concern.

IDD, although previously prevalent in Europe, North America and Australia, is now seen predominantly in developing countries. The greatest prevalence tends to be in mountainous areas such as the Andes and the Himalayas and in plateau areas far from the sea. For example, an investigation carried out by the author in the Ukinga Highlands of Tanzania revealed that 75 percent of the population had some enlargement of the thyroid.

Fluorine

Fluorine is a mineral element found mainly in the teeth and skeleton. Traces of fluorine in the teeth help to protect them against decay. Fluorides consumed during childhood become a part of the dental enamel and make it more resistant to the weak organic acids formed from foods that adhere to or get stuck between the teeth. This strengthening greatly reduces the chances of decay or caries developing in the teeth. Some studies have suggested that fluoride may also help strengthen bone, particularly later in life, and may thus inhibit the development of osteoporosis.

Dietary sources

The main source of fluorine for most human beings is the water they drink. If the water has a fluorine content of about one part per million (1 ppm), then it will supply adequate fluorine for the teeth. However, many water supplies contain much less than this amount. Fluorine is present in bone; consequently small fish that are consumed whole are a good source. Tea has a high fluorine content. Few other foods contain much fluorine.

Deficiency

If the fluoride content of drinking-water in any locality is below 0.5 ppm, dental caries will probably be much more prevalent than where the concentration is higher.

The recommended level of fluoride in water is between 0.8 and 1.2 ppm. In some countries or localities where the content of fluorine in the water is less than 1 ppm, it has now become the practice to add fluoride to the water supply. This practice is strongly recommended, but it is only practicable for large piped-water supplies; in some developing countries where most people do not have piped water, it is not feasible. The addition of fluoride to toothpaste also helps reduce dental caries. Fluorine does not totally prevent dental caries, but it can reduce the incidence by 60 to 70 percent.

Excess

An excessively high intake of fluoride causes a condition known as dental fluorosis, in which the teeth become mottled. It is usually caused by consuming excessive fluoride in water supplies that have high fluoride levels. In some parts of Africa and Asia, natural waters contain over 4 ppm of fluoride. Very high fluorine intakes also cause bone changes with sclerosis (added bone density), calcification of muscle insertions and exostoses. A survey carried out by the author in Tanzania revealed a high incidence of fluorotic bone changes (as shown by X-ray) in older subjects who normally drank water containing over 6 ppm of fluoride. Similar findings have been well described in India. Skeletal fluorosis can cause severe pain and serious bone abnormalities.

Zinc

Zinc is an essential element in human nutrition, and its importance to human health has received much recent attention. Zinc is present in many important enzymes essential for metabolism. The body of a healthy human adult contains 2 to 3 g of zinc and requires around 15 mg of dietary zinc per day. Most of the zinc in the body is in the skeleton, but other tissues (such as the skin and hair) and some organs (particularly the prostate) have relatively high concentrations.

Dietary sources

Zinc is present in most foods both of vegetable and of animal origin, but the richest sources tend to be protein-rich foods such as meat, seafoods and eggs. In developing countries, however, where most people consume relatively small amounts of these foods, most zinc comes from cereal grains and legumes.

Absorption and utilization

As with iron, absorption of zinc from the diet is inhibited by food constituents such as phytates, oxalate and tannins. No simple tests of human zinc status are known, however. Indicators used include evidence of low dietary intake, low blood serum zinc levels and low quantities of zinc in hair specimens.

Much research on this mineral has been undertaken in the last two decades, and a great deal of knowledge concerning zinc metabolism and zinc deficiency in animals and humans has been gathered. Nonetheless, there is little evidence to suggest that zinc deficiency is an important public health problem for large numbers of people in any country, industrialized or developing. However, research now under way may show that poor zinc status is responsible for poor growth, reduced appetite and other conditions; in this way zinc deficiency may contribute especially to what is now called protein-energy malnutrition (PEM).

Zinc deficiency is responsible for a very rare congenital disease known as acrodermatitis enteropathica. It responds to zinc therapy. Some patients receiving all of their nutrients intravenously have developed skin lesions which also respond to zinc treatment. In the Near East, particularly in the Islamic Republic of Iran and Egypt, a condition has been described in which adolescent or near-adolescent boys are dwarfed and have poorly developed genitalia and delayed onset of puberty; this condition has been said to respond to zinc treatment.

Zinc deficiency has also been reported as secondary to, or as a part of, other conditions such as PEM, various malabsorption conditions, alcoholism including cirrhosis of the liver, renal disease and metabolic disorders.

Other Trace Elements

Numerous minerals are present in the human body. For most of the trace elements, besides those discussed above, there is no evidence that deficiency is responsible for major public health problems anywhere. Some of these minerals are very important in metabolism or as constituents of body tissues. Many of them have been studied, and their chemistry and biochemistry have been described. Experimental deficiencies have been produced in laboratory animals, but most human diets, even poor diets, do not appear to lead to important deficiencies. These minerals therefore are not of public health importance. Other trace elements are present in the body but do not have any known essential role. Some minerals, for example lead and mercury, are of great interest to health workers because excess intake has commonly resulted in toxic manifestations.

Cobalt, copper, magnesium, manganese and selenium deserve mention because of their important nutritional role, and lead and mercury because of their toxicity. These minerals are considered in detail in large comprehensive textbooks of nutrition.

Cobalt

Cobalt is of interest to nutritionists because it is an essential part of vitamin B_{12} (cyanocobalamin). When isolated as a crystalline substance, the vitamin was found to contain about 4 percent cobalt. However, cobalt deficiency does not play a part in the anaemia that results from vitamin B_{12} deficiency.

Copper

Copper deficiency is known to cause anaemia in cattle, but no such risk is known in adult humans. Some evidence suggests that copper deficiency leads to anaemia in premature infants, in people with severe PEM and in those maintained on parenteral nutrition. An extremely rare congenital condition known as Menkes' disease is caused by failure of copper absorption.

Magnesium

Magnesium is an essential mineral present mainly in the bones but also in most human tissues. Most diets contain adequate dietary magnesium, but under some circumstances, such as diarrhoea, severe PEM and other conditions, excessive body losses of magnesium occur. Such losses may lead to weakness and mental changes and occasionally to convulsions.

Selenium

Both deficiency and excess of selenium have been well described in livestock. In areas of China where the soil selenium, and therefore the food selenium, is low, a heart condition has been described; termed Keshan's disease, it is a serious condition affecting heart muscle. Chinese researchers believe it can be prevented by providing dietary selenium. Selenium deficiency has also been associated with certain cancers.

Lead

Lead is of great public health importance because it commonly causes toxicity. Human lead deficiency is not known. Lead poisoning is especially an urban problem and is most important in children. It may lead to neurological and mental problems and to anaemia. Excess lead intake may result from consumption of lead in the household (from lead-based paint or water pipes containing lead) and from intake of atmospheric lead (from motor vehicle emissions).

Mercury

Mercury deficiency is not known in humans. The concern is with excessively high intakes of mercury and the risks of toxicity. Fish in waters contaminated with mercury concentrate the mineral. There is a danger of toxicity in those who consume fish with high mercury content. Mercury poisoning resulting from consumption of seeds coated with a mercury-containing fungicide has been described in Asia, Latin America and the Near East. The effects include severe neurological symptoms and paralysis.

VITAMINS

Vitamins are organic substances present in minute amounts in foodstuffs and necessary for metabolism. They are grouped together not because they are chemically related or have similar physiological functions, but because, as their name implies, they are vital factors in the diet and because they were all discovered in connection with the diseases resulting from their deficiency. Moreover, they do not fit into the other nutrient categories (carbohydrates, fats, protein and minerals or trace metals).

When vitamins were first being classified, each was named after a letter of the alphabet. Subsequently, there has been a tendency to drop the letters in favour of chemical names. The use of the chemical name is justified when the vitamin has a known chemical formula, as with the main vitamins of the B group. Nevertheless, it is advantageous to include certain vitamins under group headings, even if they are not chemically related, since they do tend to occur in the same foodstuffs.

Here only vitamin A, five of the B vitamins (thiamine, riboflavin, niacin, vitamin B_{12} and folic acid), vitamin C and vitamin D are described in detail. Other vitamins known to be vital to health include pantothenic acid (of which a deficiency may cause the burning feet syndrome mentioned below), biotin (vitamin H), para-aminobenzoic acid, choline, vitamin E and vitamin K: (antihaemorrhagic vitamin). These vitamins are not described in detail here for one or more of the following reasons:

— deficiency is not known to occur under natural conditions in humans;
— deficiency is extremely rare even in grossly abnormal diets;
— lack of the vitamin results in disease only if it follows some other disease process that is adequately described in textbooks of general medicine;
— the role of the vitamin in human nutrition has not yet been elucidated.

None of the vitamins omitted from discussion is important from the point of view of workers studying nutrition as community health problems in most developing countries. Those wishing to learn more about these vitamins are referred to textbooks of general medicine or more detailed textbooks of nutrition.

Vitamin A (Retinol)

Vitamin A was discovered in 1913 when research workers found that certain laboratory animals stopped growing when lard (made from pork fat) was the only form of fat present in their diet, whereas when butter was supplied instead of lard (with the diet remaining otherwise the same) the animals grew and thrived. Further animal experiments showed that egg yolk and cod-liver oil contained the same vital food factor, which was named vitamin A.

It was later established that many vegetable products had the same nutritional properties as the vitamin A in butter; they were found to contain a yellow pigment called carotene, some of which can be converted to vitamin A in the human body.

Properties

Retinol is the main form of vitamin A in human diets. (Retinol is the chemical name of the alcohol derivative, and it is used as the reference standard.) In its pure crystalline form, retinol is a very pale yellow-green substance. It is soluble in fat but insoluble in water, and it is found only in animal products. Other forms of vitamin A exist, but they have somewhat different molecular configurations and less biological activity than retinol, and they are not important in human diets.

Carotenes, which act as provitamins or precursors of vitamin A, are yellow substances that occur widely in plant substances. In some foodstuffs their colour may be masked by the green plant pigment chlorophyll, which often occurs in close association with carotenes. There are several different carotenes. One of these, beta-carotene, is the most important source of vitamin A in the diets of most people living in non-industrialized countries. The other carotenes, or carotenoids, have little or no nutritional importance for humans. In the past, food analyses have often failed to distinguish beta-carotene from other carotenes.

Vitamin A is an important component of the visual purple of the retina of the eye, and if vitamin A is deficient, the ability to see in dim light is reduced. This condition is called night blindness. The biochemical basis for the other lesions of vitamin A deficiency has not been fully explained. The main change, in pathological terms, is a keratinizing metaplasia which is seen on various epithelial surfaces. Vitamin A appears to be necessary for the protection of surface tissue.

Several studies have shown that adequate vitamin A status reduces infant and child mortality in certain populations. Vitamin A supplementation reduces case fatality rates from measles. In other illnesses such as diarrhoea and respiratory infections, however, there is not strong evidence that the prevalence or duration of morbidity is reduced by vitamin A dosing.

Calculalating vitamin A content in foods

1 IU retinol = 0.3 μg retinol = 0.3 RE

1 RE = 3.33 IU retinol

1 RE = 6 μg beta-carotene

Since pure crystalline vitamin A, which is termed retinol alcohol, is now available, the vitamin A activity in foods is now widely expressed and measured using retinol equivalents (RE) rather than the international units (IU) previously used. One IU of vitamin A is equivalent to 0.3 1 μg retinol.

Humans obtain vitamin A in food either as preformed vitamin A (retinol) or as carotenes which can be converted to retinol in the body. Beta-carotene is the most important in human diets and is better converted to retinol than other carotenes. It has been determined that six molecules of beta-carotene are needed to produce one molecule of retinol; thus it takes 6 μg of carotene to make 1 μg of retinol, or 1 RE.

Dietary sources

Vitamin A itself is found only in animal products; the main sources are butter, eggs, milk, meat (especially liver) and some fish. However, most people in developing countries rely mainly on beta-carotene for their supply of vitamin A. Carotene is contained in many plant foods. Dark green leaves such as those of amaranth, spinach, sweet potato and cassava are much richer sources than paler leaves such as those of cabbage and lettuce. Various pigmented fruits and vegetables, such as mangoes, papayas and tomatoes, contain useful quantities. Carotene is also present in yellow varieties of sweet potatoes and in yellow vegetables such as pumpkins. Carrots are rich sources. Yellow maize is the only cereal that contains carotene. In West Africa much carotene is obtained from red palm oil, which is widely used in cooking. The cultivation of the very valuable oil palm has spread to other tropical regions. In Malaysia it is widely cultivated as a cash crop, but its products are mainly exported rather than consumed locally.

Both carotene and vitamin A withstand ordinary cooking temperatures fairly well. However, a considerable amount of carotene is lost when green leaves and other foods are dried in the sun. Sun-drying is a traditional method of preserving wild leaves and vegetables often used in arid regions. Since serious disease from vitamin A deficiency is common in these areas, it is important that other methods of preservation be established.

Absorption and utilization

The conversion of beta-carotene into vitamin A takes place in the walls of the intestines. Even the most efficient intestine can absorb and convert only a portion of the beta-carotene in the diet; therefore 6 mg of beta-carotene in food is equivalent to about 1 mg of retinol. If no animal products are consumed and the body must rely entirely on carotene for its vitamin A, consumption of carotene must be great enough to achieve the required vitamin A level.

Carotene is poorly utilized when the diet has a low fat content, and diets deficient in vitamin A are often deficient in fat. Intestinal diseases such as dysentery, coeliac disease and sprue limit the absorption of vitamin A and the conversion of carotene. Malabsorption syndromes and infections with common intestinal parasites such as roundworm, which are prevalent in the tropics, may also reduce the ability of the body to convert carotene into vitamin A. Bile salts are essential for the absorption of vitamin A and carotene, so persons with obstruction of the bile duct are likely to become deficient in vitamin A. Even in ideal circumstances, infants and young children do not convert carotene to vitamin A as readily as adults do.

The liver acts as the main store of vitamin A in the human and most other vertebrates, which is why fish-iiver oils have a high content of this vitamin. Retinol is transported from the liver to other sites in the body by a specific carrier protein called retinol binding protein (RBP). Protein deficiency may influence vitamin A status by reducing the synthesis of RBP.

Storage in the body

The storage of vitamin A in the liver is important, for in many tropical diets foods containing vitamin A and carotene are available seasonally. If these foods are eaten in fairly large quantities when available (usually during the wet season), a store can be built up which will help tide the person over the dry season, or at least part of it. The short mango season provides an excellent opportunity for youngsters, who may happily spend their leisure hours foraging for this fruit, to replenish the vitamin A stored in the liver.

Toxicity

If taken in excess, vitamin A has undesirable toxic effects. The most marked toxic effect is an irregular thickening of some long bones, usually accompanied by headache, vomiting, liver enlargement, skin changes and hair loss. Cases of vitamin A toxicity from dietary sources are rare, but toxicity can be a serious problem with supplemental doses of vitamin A. A high risk of birth defects is associated with supplements given before or during pregnancy.

Human requirements

The intake recommended by FAO and the World Health Organization (WHO) is 750 μg of retinol per day for adults; lactating mothers need 50 percent more, and children and infants less. It should be noted that these figures are based upon mixed diets containing both vitamin A and carotene. When the diet is entirely of vegetable origin, larger amounts of carotene are suggested, because the conversion from carotene to retinol is not very efficient.

Deficiency

Deficiency results in pathological drying of the eye, leading to xerophthalmia and sometimes keratomalacia and blindness. Other epithelial tissues may be affected; in the skin, follicular keratosis may be the result.

Thiamine (Vitamin B_1)

During the 1890s in Java, Indonesia, Christiaan Eijkman of the Netherlands noticed that when his chickens were fed on the same diet as that normally consumed by his beriberi patients, they developed weakness in their legs and other signs somewhat similar to those of beriberi. The diet of the beriberi patients consisted mainly of highly milled and refined rice (known as polished rice). When Eijkman changed the diet of the chickens to whole-grain rice, they began to recover. He showed that there was a substance in the outer layers and germ of the rice grain that protected the chickens from the disease.

Researchers continued to work on isolating the cause of the different effects of diets of polished and whole-grain rice, but despite many attempts it was not until 1926 that vitamin B_1 was finally isolated in crystalline form. It was synthesized ten years later, and now the term thiamine is used, rather than vitamin B_1.

Properties

Thiamine is one of the most unstable vitamins. It has a rather loosely bound structure and decomposes readily in an alkaline medium. Thiamine is highly soluble in water. It resists temperatures of up to 100°C, but it tends to be destroyed if heated further (e.g. if fried in a hot pan or cooked under pressure).

Much research has been carried out on the physiological effects and biochemical properties of thiamine. It has been shown that thiamine has a very important role in carbohydrate metabolism in humans. It is utilized in the complicated mechanism of the breakdown, or oxidation, of carbohydrate and the metabolism of pyruvic acid.

The energy used by the nervous system is derived entirely from carbohydrate, and a deficiency of thiamine blocks the final utilization of carbohydrate, leading to a shortage of energy and lesions of the nervous tissues and brain. Because thiamine is involved in carbohydrate metabolism, a person whose main supply of energy comes from carbohydrates is more likely to develop signs of thiamine deficiency if his or her food intake is decreased. For this reason, thiamine requirements are sometimes expressed in relation to intake of carbohydrate. Thiamine has been synthesized in pure form and is now measured in milligrams.

Dietary sources

Thiamine is widely distributed in foods of both vegetable and animal origin. The richest sources are cereal grains and pulses. Green vegetables, fish, meat, fruit and milk all contain useful quantities. In seeds such as cereals, the thiamine is present mainly in the germ and in the outer coats; thus much can be lost during milling. Bran of rice, wheat and other cereals tends to be naturally rich in thiamine. Yeasts are also rich sources. Root crops are poor sources. Cassava, for example, contains only about the same low quantity as polished, highly milled rice. It is surprising that beriberi is not common among the many people in Africa, Asia and Latin America whose staple food is cassava.

Because it is very soluble in water, thiamine is liable to be lost from food that is washed too thoroughly or cooked in excess water that is afterwards discarded. For people on a rice diet, it is especially important to prepare rice with just the amount of water that will be absorbed in cooking, or to use water that is left over in soups or stews, for this water will contain thiamine and other nutrients.

Cereals and pulses maintain their thiamine for a year or more if they are stored well, but if they are attacked by bacteria, insects or moulds the content of thiamine gradually diminishes.

Absorption and storage in the body

Thiamine is easily absorbed from the intestinal tract, but little is stored in the body. Experimental evidence indicates that humans can store only enough for about six weeks. The liver, heart and brain have a higher concentration than the muscles and other organs. A person with a high intake of thiamine soon begins to excrete increased quantities in the urine. The total amount in the body is about 25 mg.

Human requirements

A daily intake of 1 mg of thiamine is sufficient for a moderately active man and 0.8 mg for a moderately active woman. Pregnant and lactating women may need more. FAO and WHO recommend an intake of 0.4 mg per 1 000 kcal for most persons.

Deficiency

Deficiency of thiamine leads to the disease beriberi, which in advanced forms produces paralysis of the limbs. In alcoholics thiamine deficiency leads to a condition termed Wernicke-Korsakoff syndrome.

Riboflavin (Vitamin B_2)

Early work on the properties of vitamins in yeast and other foodstuffs showed that antineuritic factors were destroyed by excessive heat, but that a growth-promoting factor was not destroyed in this way. This factor, riboflavin, was later isolated from the heat-resistant portion. It was synthesized in 1935.

Properties

Riboflavin is a yellow crystalline substance. It is much less soluble in water and more heat resistant than thiamine. The vitamin is sensitive to sunlight, so milk, for example, if left exposed may lose considerable quantities of riboflavin. Riboflavin acts as a coenzyme involved with tissue oxidation. It is measured in milligrams.

Dietary sources

The richest sources of riboflavin are milk and its non-fat products. Green vegetables, meat (especially liver), fish and eggs contain useful quantities. However, the main sources in most Asian, African and Latin American diets, which do not contain much of the above products, are usually cereal grains and pulses. As with thiamine, the quantity of riboflavin present is much reduced by milling. Starchy foods such as cassava, plantains, yams and sweet potatoes are poor sources.

Human requirements

Approximately 1.5 mg of riboflavin per day is an ample amount for an average adult, but rather more may be desirable during pregnancy and lactation. The FAO/WHO requirement is 0.55 mg per 1 000 kcal in the diet.

Deficiency

In humans a deficiency of riboflavin is termed ariboflavinosis. It may be characterized by painful cracking of the lips (cheilosis) and at the corners of the mouth (angular stomatitis). Ariboflavinosis is common in most countries but is not life threatening.

Niacin

As the history of thiamine is linked with the disease beriberi, so the history of niacin is closely linked with the disease pellagra. beriberi is associated with the East and a rice diet, and pellagra with the West and a maize diet. Pellagra was first attributed to a poor diet over 200 years ago by the Spanish physician Gaspar Casal. At first, it was believed that pellagra might be caused by a protein deficiency, because the disease could be cured by some diets rich in protein. Later it was shown that a liver extract almost devoid of protein could cure pellagra. In 1926 J. Goldberger, in the United States, demonstrated that yeast extract contained a pellagra-preventing (PP) non-protein substance. In 1937 niacinamide or nicotinamide (nicotinic acid amide) was isolated, and this was found to cure a pellagra-like disease of dogs known as black tongue.

Because pellagra was found mainly in those whose staple diet was maize, it was assumed that maize was particularly poor in niacin. It has since been shown that white bread contains much less niacin than maize. However, the niacin in maize is not fully available because it is in a bound form.

The discovery that the amino acid tryptophan prevents pellagra in experimental animals, just as niacin does, complicated the picture until it was shown that tryptophan is converted to niacin in the human body. This work vindicated and explained the early theories that protein could prevent pellagra. The fact that zein, the main protein in maize, is very deficient in the amino acid tryptophan further explains the relationship between maize and pellagra. It has also been shown that a high intake of leucine, as occurs with diets based on sorghum, interferes with tryptophan and niacin metabolism and may cause pellagra.

Properties

Niacin, a derivative of pyridine, is a white crystalline substance, soluble in water and extremely stable. It has been synthesized. The main role of niacin in the body is in tissue oxidation. The vitamin occurs in two forms, nicotinic acid and nicotinamide (niacinamide). Niacin is measured in milligrams.

Dietary sources

Niacin is widely distributed in foods of both animal and vegetable origin. Particularly good sources are meat (especially liver), groundnuts and cereal bran or germ. As for other B vitamins, the main source of supply tends to be the staple food. whole-grain or lightly milled cereals, although not rich in niacin, contain much more than highly milled cereal grains. Starchy roots, plantains and milk are poor sources. Beans, peas and other pulses contain amounts similar to those in most cereals.

Although the niacin in maize does not seem to be fully utilizable, treatment of maize with alkalis such as lime water, which is a traditional method of processing in Mexico and elsewhere, makes the niacin much more available. Cooking, preservation and storage of food cause little loss of niacin.

Human requirements

An adequate quantity for any person is 20 mg per day. Niacin requirements are affected by the amount of tryptophan containing protein consumed and also by the staple diet (i.e. whether it is maize-based or not). The FAO/WHO requirement is 6.6 mg per 1000 kcal in the diet.

Deficiency

A deficiency of niacin leads to pellagra, the "disease of the three Ds": dermatitis, diarrhoea and dementia. Initially manifested as skin trouble, pellagra, if untreated, can continue for many years, growing steadily worse.

Vitamin B_{12} (Cyanocobalamin)

Pernicious anaemia, so named because it invariably used to be fatal, was known for many years before its cause was determined. In 1926 it was found that patients improved if they ate raw liver. This finding led to the preparation of liver extracts, which controlled the disease when given by injection. In 1948 scientists isolated from liver a substance they called vitamin B_{12}. When given in very small quantities by injection, this substance was effective in the treatment of pernicious anaemia.

Properties

Vitamin B_{12} is a red crystalline substance containing the metal cobalt. It is necessary for the production of healthy red blood cells. A small addition of vitamin B_{12} or of foods rich in this substance to the diet of experimental animals results in increased growth. It is measured in micrograms.

Dietary sources

Vitamin B_{12} is present only in foods of animal origin. It can also be synthesized by many bacteria. Herbivorous animals such as cattle get their vitamin B_{12} from the action of bacteria on vegetable matter in their rumen. Humans apparently do not obtain vitamin B_{12} by bacterial action in their digestive tracts. However, fermented vegetable products may provide vitamin B_{12} in human diets.

Human requirements

The human daily requirement of this vitamin is quite small, probably around 3μg for adults. Diets containing smaller amounts do not seem to lead to disease.

Deficiency

Pernicious anaemia is not caused by a dietary deficiency of vitamin B_{12} but by an inability of the subject to utilize the vitamin B_{12} in the diet because of a lack of an intrinsic factor in gastric secretions. It may be that an autoimmune reaction limits absorption of vitamin B_{12} In pernicious anaemia the red blood cells are macrocytic (larger than normal) and the bone marrow contains many abnormal cells called megaloblasts. This macrocytic or megaloblastic anaemia is accompanied by a lack of hydrochloric acid in the stomach (achlorhydria). Later, serious changes take place in the spinal cord, leading to progressive neurological symptoms. If left untreated, the patient dies.

Treatment consists of injection of large doses of vitamin B_{12} When the blood characteristics have returned to normal, the patient can usually be maintained in good health if given one injection of 250 mg of vitamin B_{12} every two to four weeks.

Vitamin B_{12} will also cure the anaemia accompanying the disease sprue. This is a tropical condition in which the absorption of vitamin B_{12}, folic acid and other nutrients is impaired.

The tapeworm Diphyllobothrium latum, acquired from eating raw or undercooked fish, lives in the intestines and has a propensity for removing vitamin B_{12} from the food of its host. This results in the development in humans of a megaloblastic anaemia which can be cured by injection of vitamin B_{12} and treatment to rid the patient of the tapeworm.

Some medicines interfere with absorption of vitamin B_{12}.

Except in the above conditions deficiency of vitamin B_{12} is likely to occur only in those on a vegetarian diet. Deficiency causes macrocytic anaemia and may produce neurological symptoms; however, even though strict vegetarians get very little vitamin B_{12} in their diet, it appears that macrocytic anaemia due to vitamin B_{12}deficiency is not prevalent and is not a major public health problem.

Folic Acid or Folates

In 1929 Lucy Wills first described a macrocytic anaemia (an anaemia in which the red cells are abnormally large) commonly found among pregnant women in India. This condition responded to certain yeast preparations even though it did not respond to iron or any known vitamin. The substance present in the yeast extract that cured the macrocytic anaemia was at first called "Wills' factor". In 1946 a substance called folic acid, which had been isolated from spinach leaves, was found to have the same effect.

Properties

Folic acid is the group name (also termed folates or folacin) given to a number of yellow crystalline compounds related to pteroglutamic acid. Folic acid is involved in amino acid metabolism. The folic acid in foodstuffs is easily destroyed by cooking. It is measured in milligrams.

\Dietary sources

The richest sources are dark green leaves, liver and kidney. Other vegetables and meats contain smaller amounts.

Human requirements

The recommended daily intake for adults has been set at 400 μg in the United States.

Deficiency

Folate deficiency is most commonly due to poor diets, but it may result from malabsorption. It can be induced by medicines such as those used in treatment of epilepsy. A deficiency leads to the development of macrocytic anaemia. Anaemia resulting from folate deficiency is the second most common type of nutritional anaemia, after iron deficiency.

Folic acid deficiency during pregnancy has been found to cause neural tube defects in newborn babies. The role of folic acid in prevention of ischaemic heart disease has also recently received increased attention.

The main therapeutic use of folic acid is in the treatment of nutritional macrocytic or megaloblastic anaemias of pregnancy and infancy and for the prevention of neural tube defects. A dose of 5 to 10 mg daily is recommended for an adult.

Although administration of folic acid will improve the blood picture of persons with pernicious anaemia, the nervous system symptoms will neither be prevented nor improved by it. For this reason, folic acid should never be used in the treatment of pernicious anaemia, except in conjunction with vitamin B_{12}.

Vitamin C (Ascorbic Acid)

The discovery of vitamin C is associated with scurvy, which was first recorded by seafarers who made prolonged journeys. In 1497 Vasco da Gama described scurvy among the crew of his historical voyage from Europe around the southern tip of Africa to India; more than half the crew died of the disease. It gradually became apparent that scurvy occurred only in persons who ate no fresh food. It was not until 1747, however, that James Lind of Scotland demonstrated that scurvy could be prevented or cured by the consumption of citrus fruit. This finding led to the introduction of fresh food, especially citrus products, to the rations of seafarers. Subsequently scurvy became much less common.

In the nineteenth century, however, scurvy began to occur among infants receiving the newly introduced preserved milk instead of breastmilk or fresh cows' milk. The preserved milk contained adequate carbohydrate, fat, protein and minerals, but the heat used in its processing destroyed the vitamin C, so the infants got scurvy.

Later vitamin C was found to be ascorbic acid, which had already been identified.

Properties

Ascorbic acid is a white crystalline substance that is highly soluble in water. It tends to be easily oxidized. It is not affected by light, but it is destroyed by excessive heat, especially when in an alkaline solution. It is a powerful reducing agent and antioxidant and can therefore reduce the harmful action of free radicals. It is also important in enhancing the absorption of the non-haem iron in foods of vegetable origin.

Ascorbic acid is necessary for the proper formation and maintenance of intercellular material, particularly collagen. In simple terms, it is essential for producing part of the substance that binds cells together, as cement binds bricks together. In a person suffering from ascorbic acid deficiency, the endothelial cells of the capillaries lack normal solidification. They are therefore fragile, and haemorrhages take place. Similarly, the dentine of the teeth and the osteoid tissue of the bone are improperly formed. This cell-binding property also explains the poor scar formation and slow healing of wounds manifest in persons deficient in ascorbic acid.

It is a common belief, claimed also by some scientists, that very large doses of vitamin C both prevent and reduce symptoms of the common cold (coryza). This claim has not been verified. One large study did suggest a modest reduction in the severity of cold symptoms in those taking vitamin C medicinally, but the vitamin Did not prevent colds from occurring. It is not advisable to take very large doses of medicinal vitamin C for long periods of time.

Dietary sources

The main sources of vitamin C in most diets are fruits, vegetables and various leaves. In pastoral tribes milk is often the main source. Plantains and bananas are the only common staple foods containing fair quantities of vitamin C. Dark green leaves such as amaranth and spinach contain far more than pale leaves such as cabbage and lettuce. Root vegetables and potatoes contain small but useful quantities. Young maize provides some ascorbic acid, as do sprouted cereals and pulses. Animal products such as meat, fish, milk and eggs contain small quantities.

As vitamin C is easily destroyed by heat, prolonged cooking of any food may destroy much of the vitamin C present. Ascorbic acid is measured in milligrams of the pure vitamin.

Human requirements

Opinions regarding human requirements differ widely. It seems clear that as much as 75 mg per day is necessary if the body is to remain fully saturated with vitamin C. However, individuals appear to remain healthy on intakes as low as 10 mg per day. A recommendation of 25 mg for an adult, 30 mg for adolescents, 35 mg during pregnancy and 45 mg during lactation seems to be a reasonable compromise.

Deficiency

Scurvy and the other clinical manifestations of vitamin C deficiency are common in low-income countries. Scurvy is not now a prevalent disease.

Outbreaks have occurred in famine areas and recently in several refugee camps in Africa.

In its early stages vitamin C deficiency may lead to bleeding gums and slow healing of wounds.

Vitamin D

Vitamin D is associated with prevention of the disease rickets and its adult counterpart osteomalacia (softening of the bones). Rickets was for many years suspected to be a nutritional deficiency disease, and in certain parts of the world cod-liver oil was used in its treatment. However, it was not until 1919 that Sir Edward Mellanby, using puppies, demonstrated conclusively that the disease was indeed of nutritional origin and that it responded to vitamin D in cod-liver oil. Later it was proved that action of sunlight on the skin leads to the production of the vitamin D used by humans.

Properties

A number of compounds, all sterols closely related to cholesterol, possess antirachitic properties. It was found that certain sterols that did not have these properties became antirachitic when acted upon by ultraviolet light. The two important activated sterols are vitamin D_2 (ergocalciferol) and vitamin D_3 (cholecalciferol).

In human beings, when the skin is exposed to the ultraviolet rays of sunlight, a sterol compound is activated to form vitamin D, which is then available to the body and which has exactly the same function as vitamin D taken in the diet. Dietary vitamin D is only absorbed from the gut in the presence of bile.

The function of vitamin D in the body is to allow the proper absorption of calcium. Vitamin D formed in the skin or absorbed from food acts like a hormone in influencing calcium metabolism. Rickets and osteomalacia, though diseases in which calcium is deficient in certain tissues, are caused not by calcium deficiency in the diet but by a lack of vitamin D which would allow proper utilization of the calcium in the diet.

Vitamin D is often expressed in international units; 1 IU is equivalent to 0.025 μg of vitamin D3.

Dietary sources

Vitamin D occurs naturally only in the fat in certain animal products. Eggs, cheese, milk and butter are good sources in normal diets. Meat and fish contribute small quantities. Fish-liver oils are very rich. Cereals, vegetables and fruit contain no vitamin D.

Storage in the body

The body has a considerable capacity to store vitamin D in fatty tissue and in the liver. An adequate store is important in a pregnant woman, to avoid predisposition to rickets in the child.

Human requirements

It is not possible to define human dietary requirements, because the vitamin is obtained both by eating foods containing vitamin D and by the action of sunlight on the skin. There is no need for adults to have any vitamin D in their diets, provided they are adequately exposed to sunlight, and many children in Asia, Latin America and Africa survive in good health on a diet almost completely devoid of vitamin D. It has been shown that fish-liver oil containing 400 IU (10 μg) of vitamin D will prevent the occurrence of rickets in infants or children not exposed to sunlight. This amount seems to be a safe allowance.

Deficiency

As vitamin D is produced in humans by the action of the sun on the skin, deficiency is not common in tropical countries, although synthesis of vitamin D may possibly be reduced in darkly pigmented skin. Rickets and osteomalacia are seen sporadically but are more common in areas where tradition or religion keeps women and children indoors. Many cases have been reported from Yemen and Ethiopia. The conditions are manifested mainly by skeletal changes.

Toxicity

Like other fat-soluble vitamins, vitamin D taken in excess in the diet is not well excreted. Consumption of large doses, which has most commonly resulted from overdosing of children with fish-liver oil preparations, can lead to toxicity. Overdosing may lead to hypercalcaemia, diagnosed from high levels of calcium in the blood. Toxicity usually begins with loss of appetite and weight, which may be followed by mental disorientation and finally by kidney failure. Fatalities have been recorded.

Other Vitamins

The two fat-soluble vitamins (A and D) and the six water-soluble vitamins (thiamine, riboflavin, niacin, vitamin B_{12} folates and vitamin C) have been described in some detail because these are the vitamins most likely to be deficient and to be of public health importance in non-industrialized countries. Five other vitamins, although vital to human health, are not very commonly deficient in human diets and so are of less public health importance. These are vitamin B_6, biotin, pantothenic acid, vitamin E and vitamin K.

Vitamin B_6 (pyridoxine)

Vitamin B_6 is a water-soluble vitamin widely present in foods of both animal and vegetable origin. It is important as a coenzyme in many metabolic processes. Primary dietary deficiency is extremely rare, but vitamin B_6 deficiency became common in tuberculosis patients treated with the drug isoniazid. The patients developed neurological signs and some times also anaemia and dermatosis. Now it is common to provide 10 mg of vitamin B_6 by mouth daily to those receiving large doses of isoniazid. Vitamin B_6 is relatively expensive, however, and the routine administration of vitamin B_6 to patients receiving isoniazid increases the cost of treatment of tuberculosis.

Biotin

Biotin is another water-soluble vitamin of the B complex group. It is found widely in food, and deficiency in humans is extremely rare. The vitamin is very important, however, in physiological and biochemical metabolic processes. Avidin in uncooked egg white prevents absorption of biotin in animals and humans. Rats fed egg white as their only source of protein become thin and wasted and develop neuropathies and dermatitis. Biotin deficiency has been reported in a very few cases, in people consuming mainly egg white and in a few intravenously-fed patients with some special forms of malabsorption.

Pantothenic acid

Pantothenic acid, a water-soluble vitamin, is present in adequate amounts in almost all human diets. It has important biochemical functions in various enzyme reactions, but deficiency in humans is very rare. A neurological condition described as burning feet syndrome, reported in prisoners of war held by the Japanese between 1942 and 1945, was ascribed to a deficiency of this vitamin.

Vitamin E (tocopherol)

Vitamin E, a fat-soluble vitamin, is obtained by humans mainly from vegetable oils and whole-grain cereals. It has been termed the "anti-sterility vitamin" or even the "sex vitamin" because rats fed on tocopherol-deficient diets cannot reproduce: males develop abnormalities in the testicles and females abort spontaneously.

Because of its relationship to fertility and to many conditions in animals, vitamin E is widely self-prescribed and is not uncommonly recommended by physicians for a variety of human ills. However, true deficiency is probably rare; it occurs mainly in association with severe malabsorption states (when fat is poorly absorbed), in genetic anaemias [including glucose-6-phosphatase dehydrogenase (G-6PD) deficiency] and occasionally in very low-weight babies.

Vitamin E (like vitamin C) is an antioxidant, and because of its ability to limit oxidation and to deal with damaging free radicals it is sometimes recommended as a possible preventive for both arteriosclerosis and cancer. Its presence in oils helps prevent the oxidation of unsaturated fatty acids.

Vitamin K

Vitamin K has been termed the "coagulation vitamin" because of its relationship to prothrombin and blood coagulation, and because it is successfully used to treat a bleeding condition of newborn infants (haemorrhagic disease of the newborn). Humans obtain some vitamin K from food, and some is also made by bacteria in the intestines. Newborn infants have a gut free of organisms, so they do not get vitamin K from bacterial synthesis. It is now believed that intravenously-fed or starved patients receiving broad-spectrum antibiotics that kill gut bacteria may bleed because of vitamin K deficiency. In many hospitals vitamin K is given routinely to newborn infants to prevent haemorrhagic disease.

References

Hetzel, B.S.1989. *The story of iodine deficiency: an international challenge in nutrition.* New York, USA and Oxford, UK, Oxford University Press.

Shils, M.E., Olson, J.A. & Shike, M. 1994. *Modern nutrition in health and disease.* Philadelphia, Pennsylvania, USA, Lea and Febiger. 8th ed.

Souci, S.W., Fachmann, W. & Kraut, H. 1989. *Food composition and nutrition tables 1989/90.* Stuttgart, Germany, Wissenschaftliche, Verlagsgesellschaft.

UN ACC/SCN. 1991b. *Managing successful nutrition programmes.* Nutrition Policy Discussion Paper No.8. Geneva, Switzerland.

3

Nutrient-dense Foods

Different tools are used to assess the nutritional situation of groups of people, including families, communities and nations. Food composition tables provide a means to estimate the nutrient content of foods consumed by the population being studied. Tables of nutrient requirements or of recommended dietary allowances (RDAs) indicate either the suggested daily requirements of each important nutrient judged necessary to maintain satisfactory nutritional status or allowances intended to serve as goals for intakes of nutrients. These allowances often afford a margin of sufficiency; except for the energy allowance they are usually set somewhat above the physiological requirements of individuals. In general both suggested requirements and RDAs are designed for use by groups of persons, not by an individual subject. Assessment of the nutritional status of an individual needs to be based on determination of food consumption (translated into daily nutrient consumption using food composition tables), clinical examination, biochemical assessment, anthropometry and perhaps other tests.

Food balance sheets are used to provide data on food available nationally for the whole population. FAO assists many countries in assembling data on estimates of food production, imports, exports and other food uses to provide an estimate of food that was available in a particular year for the population of the country. If the population figures are available, then mean available foods can be calculated. Through the use of food composition tables, these can then be translated into mean nutrient availability [for example, the daily (or annual) availability of energy, protein and each of the important micronutrients] per head of population.

Thus food composition tables, estimates of nutrient requirements or dietary allowances and food balance sheets are tools used in different ways and for different purposes by persons wishing to assess the nutritional situation of groups of persons or of nations.

Nutrient Requirements and Recommended Dietary Allowances

A great deal of research has been conducted to determine the needs or requirements of human beings for different nutrients. Nutrient requirements of course vary in certain groups of people, for example in children because they have added needs for growth and in women during pregnancy and lactation. Large comprehensive textbooks discuss in detail the research leading to the best estimates of the requirements of different individuals for each nutrient.

Many countries provide recommendations regarding the amounts of each of the important nutrients that should be consumed by their populations. In many cases these provide for levels of safety and take account of variations in requirements; often, therefore, the figures are somewhat higher than minimum requirements for health.

Generally recommended dietary allowances for a country provide only guidelines for the evaluation and development of good diets for the population. It is important to understand clearly that the values presented are not requirements, since many individuals are known to consume smaller amounts than those listed and still enjoy good health. On the other hand, it is recognized that the actual requirement for any nutrient is not precisely known. Recommended dietary allowances therefore must not be considered requirements but rather levels of intake that should be entirely adequate for essentially all members of the population. This kind of dietary guidance seems appropriate in affluent countries such as the United States. It may not be appropriate in many parts of the world where there are more urgent problems and where food and money are more limiting factors for many people.

Food Balance Sheets

Many developing countries, using their own resources or with assistance from FAO or other organizations, have from time to time published food balance sheets, which are the best estimates that can be made from existing data of the total amount of food available for consumption by the human

Table 1. Average individual energy requirements and safe levels of intake for protein and iron

Sex and age group	Weight	Energy	Protein		Fat	Iron	
	(kg)	*(kcal)*	Diet A	Diet B	*(g)*	Diet 1	Diet 2
			(g)	*(g)*		*(mg)*	*(mg)*
Children							
6-12 months	8.5	950	14	14	-	21	11
1-3 years	11.5	1 350	22	13	23-52	13	7
3-5 years	15.5	1 600	26	16	27-62	14	7
5-7 years	19.0	1 820	30	19	30-71	19	10
7-10 years	25.0	1 900	34	25	32-74	23	12
Boys							
10-12 years	32.5	2 120	48	33	35-82	23	12
12-14 years	41.0	2 250	59	41	38-88	36	18
14-16years	52.5	2650	70	49	44-103	36	18
16-18years	61.5	2770	81	55	46-108	23	11
Girls[f]							
10-12years	33.5	1 905	49	34	32-74	23	11
12-14years	42.0	1 955	59	40	33-76	40	20
14-16years	49.5	2030	64	45	34-79	40	20
16-18 years	52.5	2 060	63	44	34-80	48	24

(contd...)

(contd...)

Men - active							
18 60 years	63.0	2 895	55	47	48-113	23	11
>60 years	63.0	2 020	55	47	34-79	23	11
Women - active							
Not pregnant or lactating	55.0	2 210	49	41	37-86	48	24
Pregnant	55 0	2 410	56	47	40-94	(76)	(38)
Lactating	55.0	2 710	69	59	45-105	26	13
>60 years	55.0	1 835	49	41	31-71	19	9

Table 2. Safe levels of intake for various micronutrients

Sex and	**Iodine**	**Vitamin A**	**Riboflavin**	**Niacin**	**Folate**	**Vitamin C**
age group	**(µg)**	***(µg retinol)***	***(mg)***	***(mg)***	***(µg)***	***(mg)***
Children						
6-12 months	50	350	0.5	5.4	32	20
1-3 years	70	400	0.8	9.0	50	20
3-5 years	90	400	1.0	10.5	50	20
5-7 years	90	400	1.1	12.1	76	20
7-10 years	120	400	1.3	14.5	102	20

(contd...)

(contd...)

Boys						
10 12 years	150	500	1.6	17.2	102	20
12-14years	150	600	1.7	19.1	170	30
14-16years	150	600	1.8	19.7	170	30
16-18 years	150	600	1.8	20.3	200	30
Girls						
10-12 years	150	500	1.4	15.5	102	20
12-14years	150	600	1.5	16.4	170	30
14-16years	150	550	1.5	15.8	170	30
16-18years	150	500	1.4	15.2	170	30
If pregnant	175	600	1.6	17.5	420	30
Men - active						
18-60 years	150	600	1.8	19.8	200	30
>60 years	150	600	1.8	19.8	200	30
Women - active						
Not pregnant or lactating	150	500	1.3	14.5	170	30
Pregnant	175	600	1.5	16.8	420	30
Lactating	200	850	1.7	18.2	270	30
>60 years	150	500	1.3	14.5	170	30

population in a particular year (or other period). Usually these estimates are based on the total quantities of food produced in the country, the foods imported and the changes up and down of food reserves or food stocks for the period. Deductions are made for foods such as cereals or legumes used as seeds rather than for consumption, for those used for livestock consumption (termed "animal feeds"), for those used for industrial non-food purposes (for example, fats or oils used for production of soap or for ethanol fuels) and for a waste factor. The final figures deduced are construed to represent the amount of food potentially available for consumption by the population of the country.

These figures then can be divided by the total mid-year population of the country to derive average per caput availability of food for the year, which can in turn be translated into per caput availability of nutrients using food composition tables. Total availability of energy and of other nutrients for the nation can also be calculated. These figures can then be compared with the calculated nutrient needs of the country to assess the adequacy of food availability. Most important, the data provide information on the dietary energy supply (DES), which combined with information about the distribution of food supplies allows an estimate of the number of people whose energy intakes are too low. The main limitation of DES is that it is not a direct assessment of food consumption. Food balance sheets also do not take into account age and gender factors, internal distribution differences within a country or seasonal variations in food availability.

Food balance sheets are often used to indicate a country's sufficiency and/or deficiency of food or particular nutrients. When prepared over successive years, they show trends in the country's food availability, indicating whether it is improving or declining and thereby allowing the country to institute appropriate policy to safeguard national food security and to channel agricultural production. The tables may also help the country to devise appropriate crop diversification policy to improve both agricultural income and production of nutritionally desirable foods. In addition the data indicate how much a country depends on its own food production relative to food importation and could thus contribute to the design of national food importation policy.

Food balance sheets for most poor developing countries are only very rough estimates of the food situation. The accuracy of the data used in the preparation of food balance sheets varies widely depending on the availability of good-quality data and the level of development of agrostatistics services. Generally they are much better in developed than in developing countries. Accurate census data on population are not available in many developing countries. Therefore the limitations of food balance sheets should be examined critically before the information is used in designing important agricultural, food security or economic policies in a particular country.

CEREALS

Early peoples lived mainly on foods obtained by hunting and gathering. Among the first crops to be planted and harvested were the cereal grains. Ancient civilizations flourished partly because of their abilities to produce, store and distribute these cereal grains: maize in the Americas before the arrival of Europeans; rice in the great Asian civilizations; and barley in Ethiopia and northeast Africa.

Foods with a predominantly carbohydrate content are important because they form the basis of most diets, especially for poorer people in the developing world. In the developing countries these foods usually provide 70 percent or more of the energy intake of the population. In contrast, in the United States and Europe often less than 40 percent of energy comes from carbohydrates.

Through the ages many plants from the family of grasses have been cultivated for their edible seeds; these are the cereal grains. Cereals form an important part of the diet of many people. They include maize, sorghum, millets, wheat, rice, barley, oats, teff and quinoa. A new cereal of considerable interest is triticale, a cross between wheat and rye.

Although the shape and size of the seed may be different, all cereal grains have a fairly similar structure and nutritive value; 100 µg of whole grain provides about 350 kcal, 8 to 12 µg of protein and useful amounts of calcium, iron (though phytic acid may hinder absorption) and the B vitamins (Table 3).

In their dry state cereal grains are completely lacking in vitamin C and, except for yellow maize, contain no carotene (provitamin A). For a balanced diet, cereals should be supplemented with foods rich in protein, minerals and vitamins A and C. (Vitamin D can be obtained through exposure of the skin to sunlight.)

Table 3. Content of certain nutrients in 100 g of selected cereals

Food	Energy	Protein	Fat	Calcium	Iron	Thiamine	Riboflavin	Niacin
	(kcal)	*(g)*	*(g)*	*(mg)*	(mg)	(mg)	(mg)	(mg)
Maize flour, whole	353	9.3	3.8	10	2.5	0.30	0.10	1.8
Maize flour, refined	368	9.4	1.0	3	1.3	0.26	0.08	1.0
Rice, polished	361	6.5	1.0	4	0.5	0.08	0.02	1.5
Rice, parboiled	364	6.7	1.0	7	1.2	0.20	0.08	2.6
Wheat, whole	323	12.6	1.8	36	4.0	0.30	0.07	5.0
Wheat flour, white	341	9.4	1.3	15	1.5	0.10	0.03	0.7
Millet, bulrush	341	10.4	4.0	22	3.0	0.30	0.22	1.7
Sorghum	345	10.7	3.2	26	4.5	0.34	0.15	3.3

The structure of all cereal grains consists of:

— the husk of cellulose, which has no nutritive value for humans;
— the pericarp and testa, two rather fibrous layers containing few nutrients;
— the aleurone layer,'which is rich in protein, vitamins and minerals;
— the nutrient-rich embryo or germ, consisting of the plumule and radicle attached to the grain by the scutellum;
— the endosperm, comprising more than half of the grain and consisting mainly of starch.

The embryo is the part of the grain that sprouts if the grain is planted or soaked in water. It is very rich in nutrients. Although small in size, the embryo often contains 50 percent of the thiamine, 30 percent of the riboflavin and 30 percent of the niacin of the whole grain. The aleurone and other outer coats contain 50 percent of the niacin and 35 percent of the riboflavin. The endosperm, although by far the largest part of the grain, often contains only one-third or less of the B vitamins. Compared with other parts, it is also poorer in protein and minerals, but it is the main source of energy, in the form of a complex carbohydrate, starch.

Processing

Cereal grains are subjected to many different processes during their preparation for human consumption. All of the processes have in common that they are designed to remove the fibrous layers of the grain. Some processes, however, are also intended to produce a highly refined white product consisting mainly of endosperm. Another common characteristic shared by all the processes is that they reduce the nutritional value of the grain.

Traditional methods of processing, involving the use of a pestle and mortar or stones, usually produce a cereal grain that has lost some of its outer coats but retains at least part of the germ, including the scutellum. Although very careful and prolonged processing using traditional methods can yield a highly refined product, such preparation is unusual. Light milling, similar to home pounding, also produces a product that retains most of the nutrients. Mechanization of this type has the additional advantage of taking a great burden off the woman of thehousehold, as it is usually the woman who is responsible for pounding grain.

Heavy milling to produce a highly refined product is undesirable from a nutritional point of view. Highly milled cereals such as white maize flour, polished rice and white wheat flour have lost most of the germ and outer layers and with them most of the B vitamins and some of the protein and minerals. Millers are, however, servants of the public, and the public increasingly demands products that are pure white, have a bland, neutral taste and are easily digestible. These demands led, in the first half of the twentieth century, to a vast increase in the production of highly refined cereal flours and white rice. The millers have responded to public demand by devising "improved" milling machinery that separates more and more of the nutritious parts from the grain, leaving the white endosperm.

The percentage of the original grain that remains in the flour after milling is termed the extraction rate. Thus an 85 percent extraction flour contains 85 percent (by weight) of the whole grain, 15 percent having been removed. Therefore, a high-extraction flour has lost little of the nutrients in the outer coats and germ, whereas a low-extraction flour has lost much. The advantages of low-extraction flours over high-extraction flours from the trade point of view are that they are whiter, and so more popular; they have less fat, and hence less tendency to become rancid; they have less phytic acid, which possibly means that minerals from associated foods are absorbed better; and they have better baking qualities. The disadvantages of low-extraction flours to the consumer are that they contain less B vitamins, minerals, protein and fibre than high-extraction flours.

In many countries food fashions begin among the wealthier people. As long as the new food fashion remains confined to those with a high income, it need not do much harm, for they can afford a better all-round diet which compensates for the nutrients lost to fashion. However, the white-flour fashion has spread to all levels of society, rich and poor, in many countries. In addition, highly milled rice spread across Asia quite rapidly over 80 years ago.

Where preference for white flour or highly milled rice leads to the consumption of a staple cereal rendered deficient by milling, widespread ill health could be, and has been, the result among those who do not include in their diet other foods that make up for this deficiency. Much misery, suffering and death resulted directly from the introduction of milled cereals to the people of Asia around the beginning of the twentieth century, when the disease beriberi became highly prevalent.

Increasing industrialization and urbanization in developing countries have brought with them a much greater use of bread because of its convenience for workers eating away from home.

Manufactured cereal products are being increasingly sold as baby and breakfast foods. In developing countries these products are mainly imported. They may be convenient but they are relatively expensive and have no inherent advantage, from a nutritional point of view, over local cereals prepared in a traditional manner. However, they may be highly advertised, considered as prestige foods and falsely regarded as more nutritious than local foods. Their use should be discouraged for those who cannot really afford them.

Legislation requiring millers to put additional vitamins into cereal flours exists in some countries and can be effective. This procedure does not work as well with rice because it is most commonly bought and eaten in its granular form, whereas maize and wheat and most other cereals are more often bought as flour. Attempts have been made in Asia to add vitamins in a concentrated form to a few artificial granules and then to mix these with rice. This method has not proved entirely successful, partly because one of the B vitamins, riboflavin, is yellow and lends a colour unacceptable to those who want a uniformly white product.

Maize

Maize (Zea mays) is a very important food in the Americas and much of Africa. It was first cultivated in the Americas, and it was an important food of the great Aztec and Mayan civilizations long before the arrival of Columbus and the colonizers. Seeds were brought to Europe and later to Africa, where maize is now the most important part of the diet in many areas. Maize is popular because it gives a high yield per unit area, it grows in warm and fairly dry areas (much drier than those needed for rice, although not as dry as those where sorghum and millet can be raised), it matures rapidly and it has a natural resistance to bird damage. The United States is the largest producer of maize, but much of what is grown there is used to feed domestic animals.

Nutrient content. Maize grains contain about the same amount of protein as other cereals (8 to 10 percent), but much of it is in the form of zein, a poor-quality protein containing only small amounts of lysine and tryptophan. The association noted between maize consumption and pellagra may be due in part to a deficiency of these amino acids. Whole-grain maize contains 2 mg niacin per 100 g, which is less than that in wheat or rice and about the same as the amount in oats. The niacin in maize is in a bound form and not entirely available to humans. In Mexico and some other countries maize is treated with an alkaline solution of lime which releases the niacin and helps prevent pellagra; maize treated with lime is used for making tortillas, an important food in Mexico. New varieties of maize, for example opaque-2 maize, have now been developed with an improved amino acid pattern.

Processing. Milling reduces the nutritive value of maize just as it does that of other cereals. The increased popularity and use of highly milled maize meal as opposed to traditionally ground or lightly milled maize in Africa could create a problem, since the highly milled product is deficient in B vitamins (Table 4); it is necessary to eat 600 g of highly milled maize to obtain the amount of thiamine present in 100 g of lightly milled maize.

Table 4. Effect of milling on vitamin B content of maize (mg per 100 g)

Level of processing of maize	*Thiamine*	*Riboflavin*	*Niacin*
Whole grain	0.35	0.13	2.0
Lightly milled	0.30	0.13	1.5
Highly milled (65 percent extraction)	0.05	0.03	0.6

The vitamin B constituents lost in milling may be replaced in maize meal, as in other cereal flours, by fortification. Enrichment of this kind has been effective in many countries. Legislation to ensure an adequate level of B vitamins in cereal flours may be feasible and worthwhile for more countries to adopt.

Rice

Rice, like other cereals, is a domesticated grass; wild varieties have existed for centuries in both Asia (Oryza sativa) and Africa (Oryza glaberina). Rice is a particularly important food for much of the population of China and for many other countries in Asia, where close to half the population of the world lives. It is also important in the diets of some peoples in the Near East, Africa and to a lesser extent the Americas. Much of the rice is produced in small irrigated fields or paddies in Asia, but some is grown in rain-fed areas without irrigation.

Nutrient content. The outer layers and the germ together contain nearly 80 percent of the thiamine in the rice grain. The endosperm, though constituting 90 percent of the weight of the grain, contains less than 10 percent of the thiamine. Lysine and threonine are the limiting amino acids in rice.

Processing. After harvesting, the rice seeds or grains are subjected to different milling methods. The traditional home method of pounding rice in a wooden mortar and winnowing it in a shallow tray usually results in the loss of about half of the outer layers and germ, leaving a product containing about 0.25 mg thiamine per 100 g. The procedure of milling and subsequently polishing rice, which produces the highly esteemed white rice on sale in many shops, removes nearly all the outer layers and germ and leaves a product containing only about 0.06 mg thiamine per 100 g. This amount is grossly deficient. In Asia, many poor people have a diet consisting mainly of rice for much of the year. A person eating 500 g of highly milled polished rice per day would get only 0.3 mg thiamine. The same quantity of home-pounded or lightly milled rice would provide approximately 1.25 mg thiamine, which is about the normal requirement for an average man.

Fortification is one method of adding micronutrients. Another way of providing highly milled rice that is reasonably white and yet contains adequate quantities of B vitamins is by parboiling. This process is usually done in the mill, but it can be done in the home. The paddy, or unhusked rice, is usually steamed, so that water is absorbed by the whole grain, including the endosperm. The B vitamins, which are water soluble, become evenly distributed throughout the whole grain. The paddy is dried and dehusked, and it is then ready for milling in the ordinary way. Even if it is highly milled and polished, the parboiled grain still retains the major part of its thiamine and other B vitamins.

The solubility of the B vitamins has its disadvantages. Rice that is washed too thoroughly in water loses some of the B vitamins, which are dissolved out. Similarly if rice is boiled in excess water, a considerable proportion of the B vitamins is likely to be discarded with the water after cooking. Rice should therefore be cooked in just the amount of water it will absorb. If any water is left over it should be used in a soup or stew, since it will contain valuable B vitamins which should not be wasted.

Wheat

Wheat (genus Triticum) is the most widely cultivated cereal in the world and its products are very important in human nutrition. In many parts of the world where wheat cannot be grown it is imported and is becoming an increasingly important part of the diet, especially for the urban population. However, importation of wheat, like that of all products, must be offset by adequate exports to prevent a drain on a country's foreign exchange.

Bread, usually made from wheat flour, is a popular convenience food. When purchased, it saves time and fuel for poor families. Pasta is also becoming increasingly popular in some developing countries.

Nutrient content. Wheat provides a little more protein than does rice or maize, about 11 g per 100 g. The limiting amino acid is lysine. In many industrialized countries wheat flour is fortified with B vitamins and sometimes iron and other nutrients.

Processing. Wheat is usually ground and made into flour. As with other milled cereals, the nutrient content depends on the degree of milling, i.e. the extraction rate. Low-extraction flours have lost much of their nutrients. In some developing countries where wheat is being increasingly used, the bakers have encouraged the trend towards highly refined products because white wheat flour has better baking qualities. Traders also prefer the highly milled product because it stores better. Its low fat content reduces the chances of rancidity, and its low vitamin content makes it less attractive to insects and other pests.

Millets and sorghum

Millets and sorghum are cereal grains widely grown in Africa and in some countries in Asia and Latin America. Although less widely grown than maize, rice and wheat, they are important foods. They survive drought conditions better than maize and other cereals, so they are commonly grown in areas where rainfall is low or unpredictable. They are valuable food crops because they nearly all contain a higher percentage of protein than maize and the protein is also of better quality, with a fairly high content of tryptophan. These cereals are also rich in calcium and iron. Because they tend to be ground at home and not in the mill, they are less frequently subjected to vitamin, mineral and protein losses. However, in many areas of Africa, they are being replaced by rice and maize, although they usually continue to be grown for making beer. In some parts of Asia millets are regarded as low-class foods for poor people.

Many millet and sorghum varieties have the disadvantages of susceptibility to attack by small birds and a tendency to shed their grain. Losses are often high. In some countries millets and sorghum are used mainly to feed animals.

Sorghum (Sorghum vulgare or Sorghum bicolor) is believed to have originated in Africa but is now cultivated in many countries. It is also called guinea corn or durra, and in India it is known as jowar. There are many varieties of sorghum; most grow tall and have a large inflorescence, but there are also dwarf varieties. The grain is usually large but varies in colour and shape with the type. Sorghum requires more moisture than millets but less than maize. Sorghum is a nutritious food and many varieties have a higher protein content than other cereals.

There are several species of millet. The most important in Africa are bulrush millet (Pennisetum glaucum), also called pearl millet, and finger millet (Eleusine coracana). The former, as the name implies, looks rather like a bulrush, but the inflorescence may be much longer and thicker, some times 1 x 8 cm. The inflorescence of finger millet is shaped like a rather flaccid hand. The seeds are smaller than those of bulrush millet. It is very commonly used for making beer.

Other Cereals

Oats

Oats are not important in the diets of most developing countries. The crop is grown in a few cold highland areas, where it is locally prepared and not usually milled. Oats are a good cereal containing rather more protein than maize, rice or wheat, but they also contain a considerable quantity of phytic acid which may hinder absorption of iron and calcium. Oatmeal is imported for use in porridge and is used in some manufactured infant foods.

Rye

Rye is little grown in Africa, Asia and Latin America, and even in Europe it is not an important item of the diet. It has nutritive properties similar to those of other cereals and is sometimes added to bread.

Barley

Barley is grown in some of the wheat-growing districts of Africa and in highland areas of Asia and South America. In these places it is usually consumed as a stiff porridge after home preparation. In Europe it is now used mainly for animal feeding and in the preparation of alcoholic beverages such as beer and whisky.

Triticale

This new cereal is a cross between wheat and rye. It has promise of high yields and good nutritive value. It is particularly suited to temperate climates.

Teff

Teff (Eragrostis tef) is an important cereal in Ethiopia, where it is held in special regard although it gives a relatively low yield per unit area. It is usually ground into a flour, cooked and eaten as injera, a type of baked pancake. The nutritive value of teff is similar to that of other cereal grains, except that it is richer in iron and calcium. The high consumption of teff in parts of Ethiopia may be an important reason why iron deficiency anaemia is rarely reported there.

Quinoa

Quinoa is a millet-like cereal grain which is grown in South America in the Andes, particularly in the altiplano. It grows well even where rainfall is low, soil is not very fertile and nights are very cold. As a food it has a special place in the diets of some Andean peoples.

Starches and Starchy Roots

A number of edible tubers, roots and corms form an important part of the diet of many peoples in different parts of the world. In tropical countries cassava, sweet potatoes, taro (cocoyam), yams and arrowroot are the most important foods in this class. In the cooler parts of the world the common potato is also widely grown.

These food crops are usually relatively easy to cultivate and give high yields per hectare. They contain large quantities of starch and are therefore a fairly easily obtainable source of food energy. As staple foodstuffs, however, they are inferior to cereals because they consist of about two-thirds water and have much less protein, as well as lower contents of minerals and vitamins. They usually contain less than 2 percent protein, whereas cereals contain about 10 percent. Taro and yams, however, contain up to 6 percent good-quality protein.

Cassava

Although cassava (Manihot esculenta), also known as yuca or manioc, originated in South America, it is now widely grown in many parts of Asia and Africa, mainly for its starchy tuberous roots which may grow to an enormous size. Readily established from cuttings, it will grow in poor soil, requires relatively little attention, withstands adverse weather conditions and until recently was not greatly afflicted by pests or disease. However, in some parts of Africa, notably Malawi, cassava plants in the fields have been attacked and destroyed by mealy bug.

Energy yields per hectare from cassava roots are often very high, potentially much higher than from cereals. The leaves of the plant are eaten by some societies and are nutritious. However, cassava has the great disadvantage of containing little but carbohydrate. It is especially unsuitable as the main source of energy for the infant or young child because of its low protein content. It should therefore be supplemented liberally with cereals and also with legumes or other protein-rich foods. However, in non-arid areas where the main food and nutrition problems arise from shortage of total food and deficient energy intake, cassava should be encouraged because of its high yields and other agricultural advantages.

Cassava contains less than 1 percent protein, significantly less than the 10 percent in maize and other cereals. It is not surprising, therefore, that kwashiorkor resulting from protein deficiency is much more common in young children weaned on to cassava than in those weaned on to millet or maize. Cassava also has considerably less iron and B vitamins than the cereal grains.

Cassava, particularly bitter varieties, sometimes contains a cyanogenic glucoside. This poisonous substance is present mainly near the outer coat of the tuber, so peeling cassava helps reduce the cyanide. Cassava that is soaked in water or boiled in water that is then discarded also has reduced cyanide levels. In addition, toxicity can be reduced by pounding, grating and fermentation of the cassava roots. Toxic effects tend to occur where these practices are not used. Cassava consumption has also been linked with goitre and iodine deficiency disorders.

Cassava leaves are frequently used as a green vegetable. Their nutritive value is similar to that of other dark green leaves. They are an extremely valuable source of carotene (vitamin A), vitamin C, iron and calcium. The leaves also contain some protein. To preserve the maximum quantity of vitamin C in the leaves, they should not be cooked for longer than about 20 minutes.

Cassava tubers may be eaten roasted or boiled, but more often they are sun-dried after soaking and then made into a powdery white flour. In some countries cassava is milled commercially. In some of this processing the end product is tapioca, which is mainly cassava starch. In West Africa cassava is used to make fufu (a boiled, mashed product). In some countries, for example Indonesia, cassava is regarded as a poor persons' food, and in others as a famine food.

Sweet Potatoes

Sweet potatoes originated in the Americas and are now widely grown also in tropical Africa and Asia, usually from stem cuttings. Like cassava, the irregularly shaped, variously sized tubers contain little protein. They contain some vitamin C, and the coloured varieties, especially the yellow ones, provide useful quantities of carotene (provitamin A). Sweet-potato leaves are often eaten and have properties very similar to those of cassava leaves. However, the leaves should not be picked to excess since, as with other tuber crops, this may reduce the yield of tubers.

Yams

There are innumerable varieties of yams (genus Dioscorea), some of which are indigenous to Africa, Asia and the Americas. They vary in shape, colour and size as well as in cooking quality, leaf structure and palatability. Besides the many domesticated varieties a number of wild varieties are eaten.

Yams are more extensively grown in West Africa than in East Africa. In Nigeria, for example, yams are still an important root crop despite an increase in the popularity of cassava. Yams require a warm, humid climate and soil rich in organic matter; these requirements limit their cultivation.

The proper cultivation of yams entails initial deep digging and subsequent staking of the twining vine-like plant. The work involved is more arduous than for cassava, and the yields, though high, are usually a little lower than cassava yields. Yams usually contain about twice as much protein (2 percent) as cassava, although very much less than cereals.

Taro or Cocoyams

Taro (Colocasia sp.) originated in Asia but is quite extensively grown in areas where there is fairly high rainfall spread over much of the year. It is widely grown and consumed in the Pacific islands. In Africa, taro is common in forest areas (e.g. in the Ashanti country of Ghana) and on mountain slopes where precipitation is high (e.g. Mount Kilimanjaro). Taro is often grown in association with bananas or plantains (e.g. by the Buganda) or together with oil-palms. The plant has large "elephant ear" leaves. Both the tubers and the leaves are eaten. The nutritive value of taro is similar to that of cassava. In some areas taro is being replaced by tania or new cocoyam (Xanthosoma sp.), a somewhat similar but more robust plant originally from South America, which outyields taro.

Potatoes

Potatoes were first taken to Europe from South America and became a cheap, useful, high-yielding alternative to the existing main staples, just as cassava replaced millet in parts of Africa and Asia. However, the mistake of relying almost entirely on one crop was emphasized by the great Irish potato famine of the nineteenth century: when the potato crop failed because of a blight, over one million people died and even larger numbers emigrated. Potatoes remain a very important food of people living in the Andean countries of South America. Much research on this crop has been conducted in Peru. From Europe, potatoes travelled to Africa and Asia, where they are grown in higher cool areas. If well cultivated in the right soil and climate, they can give a very high yield per hectare.

Like other starchy tubers, potatoes contain only about 2 percent protein, but the protein is of reasonably good quality. Potatoes also provide small quantities of B vitamins and minerals. They contain about 15 mg vitamin C per 100 g, but this amount is reduced in storage. The keeping quality of potatoes is not good, unless they are stored carefully.

Arrowroot

Arrowroot, which is grown in areas with adequate rainfall, is liked by certain peoples in Africa and Oceania. The nutritive value of arrowroot is similar to that of potatoes. The roots are eaten in a variety of ways, often roasted or boiled.

Other Carbohydrate Foods

Bananas and plantains

Strictly speaking, bananas and plantains should be discussed with fruits; from the nutritional point of view, however, they are more appropriately considered under starchy foods. It is difficult to differentiate among the many varieties of plantain and of banana. For the purposes of this chapter, plantains may be described as bananas that are picked green and are cooked before eating. Plantains contain more starch and less sugar than bananas, which are usually eaten raw like other fruits.

Bananas and plantains originally grew wild in damp, warm forest areas. They have probably been used as food by humans since earliest times. Bananas and plantains are now cultivated extensively in many of the humid tropical areas. Some peoples such as the Buganda in Uganda and the Wachagga in the United Republic of Tanzania depend on plantains as their main food.

A 100-g portion of green bananas or plantains provides 32 g carbohydrate (mainly as starch), 1.2 g protein, 0.3 g fat and 135 kcal. Plantains also have a high water content. Their very low protein content explains why kwashiorkor commonly occurs in young children weaned on to a mainly plantain diet. Bananas usually contain about 20 mg vitamin C and 120 mg vitamin A (as beta-carotene equivalent) per 100 g. For this reason fresh fruits and vegetables are much less important in the diet for those whose staple food is banana than for those whose staple is a cereal or root. Bananas are, however, low in their content of calcium, iron and B vitamins. As bananas supply only 80 kcal per 100 g, about 2 kg must be eaten to provide 1 500 kcal.

Plantains are usually picked while they are still green. The skin is peeled off and they are then either roasted and eaten, or, more commonly, cut up, boiled and eaten with meat, beans or other foods. Plantains are frequently sun-dried and made into a flour.

Sago

Sago (Metroxylon sp.) is almost pure starch and comes from various forms of the sago palm. The trees are widely grown in Indonesia, but sago as a food is particularly popular in certain Pacific islands. Sago has low protein content.

Sugar

Sugar, as sold in shops, is almost 100 percent sucrose and is essentially pure carbohydrate. In Africa, Asia and Latin America nearly all locally produced sugar comes from sugar cane, while in Europe and North America some comes from sugar beet.

In areas where much sugar cane is grown, the consumption of sugar or sugar-cane juice (chewed cane) is often high. In other parts of the world the consumption of sugar tends to rise with economic advancement. In the United States and the United Kingdom in 1995 about 18 percent of energy consumed came from sugar (sucrose), mainly in sweetened foods. In contrast, in many African countries less than 5 percent of energy comes from sucrose. Sugar is a good and often inexpensive source of energy and can be a valuable addition to bulky energy-deficient diets. Contrary to popular belief, customary consumption of sugar is not related to obesity, diabetes, hypertension or any other non-communicable disease. Frequent sugar consumption can be associated with dental caries when coupled with poor oral hygiene, but sucrose is no more cariogenic than other fermentable sugars.White sugar contains no vitamins, protein, fat or minerals. Many people find that its sweet taste adds to the enjoyment of eating. The yields of energy per hectare of land are very high on productive sugar estates.

Honey

From time immemorial honey has been extensively gathered in developing countries from wild hives. Now more and more hives are being kept, often in hollowed and suspended pieces of tree-trunk or in other more managed ways. The incentive to keep bees tends to be the high price of beeswax rather than just the honey.Honey has gained the false reputation of being of special nutritive value. In fact it contains only sugar (carbohydrate), water and minute traces of other nutrients. Although merely a source of energy, it has sensory value as a pleasant food for humans.

Legumes, Nuts and Oilseeds

Legumes or Pulses

Beans, peas, lentils, groundnuts and their like belong to the botanical family of Leguminosae. Their edible seeds are called legumes or pulses. Agriculturally the plants of this group have the advantage of being able to obtain nitrogen from the air and also add some to the soil, whereas most other plants take nitrogen from the soil and do not replace it. Legumes usually thrive best when they can get water early in their growth and then have a warm dry spell for ripening. They are therefore often planted at the end of the rains to ripen early in the dry season.

In Africa, Asia and Latin America the seeds are usually left on the plant to reach full maturity and are then harvested and dried. Some may be picked earlier and eaten while partly green, as in Europe and North America.

The dried seeds can be kept and stored in much the same way as cereals.

Some varieties are susceptible to attack by weevils; spending a small amount of money on insecticides to prevent this is definitely economically sound. However, care must be taken to ensure that an excess of insecticide is not applied, that the insecticide is relatively safe and that the beans are well washed before cooking.

The legumes are very important from a nutritional point of view because they are a widely available vegetable food containing good quantities of protein and B vitamins in addition to carbohydrate. Some legumes, such as groundnuts and soybeans, are also rich in oil. They usually supplement very well the predominantly carbohydrate diet based on cereals. Most legumes contain more protein than meat, but the protein is of slightly lower quality because it has less methionine. However, when pulses and cereals are eaten together at one meal they supply a protein mixture containing good quantities of all the amino acids, which improves the protein value of the diet. Legumes also contain some carotene (provitamin A) and ascorbic acid if eaten green. Similarly, dried legumes allowed to sprout before eating have good quantities of ascorbic acid. Some legumes contain antivitamins or toxins.

Unless there is a very good reason for introducing a new crop such as soybeans, it is more sensible to encourage increased production and consumption of whatever legume is already grown and popular in any area. The local people will have a taste for it, and agricultural conditions are usually suitable. It is also highly important to try to introduce beans (and other pulses) into the diet of children at an early age. Children are just as able as adults to digest beans easily.

Beans, Peas, Lentils and Grams

A wide variety of beans, peas, lentils, grams, etc. are grown and are important in the diet of people in Asia, Africa and Latin America. All three regions have indigenous legume varieties but also grow varieties that originated on other continents.

There are many kinds of beans. Haricot or kidney beans (Phaseolus vulgaris) were originally from the Americas but are now widely grown in Asia and Africa. Broad beans (Vicia faba) are more common in temperate areas. Lima beans (Phaseolus lunatus) originated from Peru but are eaten all over the tropics and subtropics. Mung beans (Phaseolus aureus), indigenous to the Indian subcontinent, are small seeds but very popular. Scarlet runner beans (Phaseolus multiflorus)are popular as a fresh vegetable in Europe and North America, but the large mature seeds are eaten dried in many countries.

Lentils (Lens esculenta) and some similar legumes often known as grams are very important in the diets of people in many developing countries. Lentils have been cultivated for food by humans for thousands of years. The plants are of small size, as are the seeds. Grams include the important pigeon pea (Cajanus cajan),chickpea (Cicer arietinum) and green gram or mung bean (Phaseolus aureus). In many South Asian countries various dhals made from these legumes form a significant part of the diet, providing important nutrients to supplement the staple food, which may be rice or wheat. In many parts of Africa both cowpeas and pigeon peas are grown and consumed. The pigeon pea is perennial and relatively drought resistant. Lathyrus sativus, another drought-resistant legume, is grown widely in India, but consumption of large amounts can lead to the severe toxic condition called lathyrism. Winged bean (Psophocarpus tetragonolobus) isanother important legume with a very high protein content (35 percent), but it is not yet widely grown.

Peas are commonly consumed as a green vegetable (fresh, canned or frozen) in Europe and North America and by more affluent people elsewhere. In developing countries the seeds are allowed to mature and are dried and consumed in the same manner as other legumes.

These legumes (excluding soybean) all have a somewhat similar nutritive value, but the mature beans are eaten in a variety of ways and have different flavours and other culinary qualities. Most legume seeds usually contain about 22 percent protein (as opposed to 1 percent in cassava roots and 10 percent in maize) and good quantities of thiamine, riboflavin and niacin; in addition they are richer in iron and calcium than most of the cereals.

The large number of other legume seeds of various shapes, colours and sizes on sale at food shops or marketplaces in almost any village or town in tropical countries is evidence of an appreciation of dietary variety and culinary finesse. Culture and local taste importantly determine how these foods are eaten.

Soybean

Soybean (Glycine max) originated in Asia, but now the main producers are the United States and Brazil. However, the soybeans produced by these countries are mainly used commercially for oil and as animal feeds. Asia still produces much of the soybeans for direct human consumption. They are not widely grown in Africa or Latin America.

Soybeans contain up to 40 percent protein, 18 percent fat and 20 percent carbohydrate. The protein is of a higher biological quality than that from other plant sources.

Soybeans, used in a wide variety of ways, are very important in the diets of the Chinese and in those of some other Asian countries. In China soybeans are made into a variety of tasty dishes which supplement the staple food of rice or other cereal. Soy products such as tofu (soybean curd) and tempeh (a fermented product) are important in Indonesian cuisine and popular elsewhere. Soybeans have not become a popular food in Africa or Latin America, where there is little local knowledge of the best methods of preparing them. People lacking experience with soybeans find them difficult to prepare and cook.

Where soybeans are grown they can be locally processed for use in the country as an enrichment of cereal flours, as an infant food or for institutional and school-feeding purposes. The oil can be exported and the protein-rich residue cake can be utilized in the country.

Groundnuts (peanuts, monkey-nuts)

The term "groundnut" is a misnomer since, although botanically a nut, the groundnut (Arachis hypogaea) is a true pulse, a member of the Leguminosae family. It originated in Brazil but is now extensively grown in warm climates around the world. It is an unusual plant in that the flower stalk bearing the ovary burrows into the ground, where a nut containing the seed or seeds develops.

Groundnuts contain much more fat than other legumes, often 45 percent, and also much more niacin (18 mg per 100 g) and thiamine, but relatively little carbohydrate (12 percent). The protein content is a little higher than that of most other pulses (27 percent). Groundnuts are an unusually nutritious food with more protein than animal meat. They are energy dense because of their oil, and they are rich in vitamins and minerals. If every child, woman and man in Africa ate a handful of groundnuts per day in addition to their normal diet, Africa would be rid of most existing malnutrition.

Groundnuts are fairly widely grown in the tropics. Farmers should produce them for home consumption as well as for cash crops, since they form a very useful addition to the primarily cereal or root diets of many poor families. They supply much-needed fat, which is high in energy and assists in the absorption of carotene as well as serving other functions. In predominantly maize diets, relatively small quantities of groundnuts, with their high content of niacin and also of protein (including the amino acid tryptophan), can prevent pellagra. When groundnuts are added to children's diets, their high protein and energy content serves to prevent protein-energy malnutrition.

However, groundnuts are often grown mainly as a cash crop even in developing countries. The world's largest producer is the United States. Groundnuts are usually utilized for oil extraction, and the residue, groundnut cake, is used for animal feed. In the United States a good proportion is consumed as peanut butter. In many countries groundnuts are consumed roasted, boiled or cooked in other ways.

Groundnuts, if damaged during harvesting or if poorly stored in damp conditions, may be attacked by the mould Aspergillus flavus. This fungus produces a poisonous toxin known as aflatoxin, which has been shown to cause liver damage in animals and to kill poultry fed on infected groundnuts. It may be toxic also for humans and may be a cause of liver cancer.

Bambara Groundnut

The bambara groundnut (Voandzeia subterranea) originated in Africa and is grown widely. It resembles the groundnut physically but is not nutritionally similar, having only 6 percent fat. Its protein content of 18 percent is a little lower than that of most other pulses, but it has about the same mineral and vitamin content as beans. Because of the lower fat content the crop is not in great demand for oil production. Therefore, instead of being sold as a cash crop, it is more often used locally for food.

Tree Nuts

Coconut

Coconut is the most important nut crop in Africa. Its origins are uncertain. The nut, being light and impervious to water, no doubt drifted across many seas to germinate on a new shore. It is now extensively cultivated. The tree that bears it is a picturesque and highly useful plant, apart from the food it provides for humans. When it is green, the nut contains about half a litre of water; this is a very refreshing and hygienic drink, but apart from a little calcium and carbohydrate, it has no nutritive value. The white flesh, however, is rich in fat.

The flesh of the coconut is usually sun-dried into copra. The oil from copra is used both for cooking and for making soap. Copra itself is used in the tropics and elsewhere as an addition to many dishes. It is an important ingredient in a variety of cuisines from Thailand to Saudi Arabia. Coconut oil has the disadvantage of containing a relatively high proportion of saturated fatty acids. The coconut sap in many countries is fermented to yield alcoholic beverages.

Cashew nut

The cashew nut is produced on a small tree that originated in dry areas of the Americas. It is widely grown in the tropics, and the nuts are mainly exported. They are rich in fat (45 percent) and contain 20 percent protein and 26 percent carbohydrate. The edible swollen stalk of the nut contains good quantities of vitamin C. Cashew nuts are a useful local food but too expensive for most people.

Oilseeds

Sesame

Sesame, or simsim (benniseed in West Africa), is grown fairly widely throughout the world and is largely used for oil extraction. The seeds, which are of various colours, contain about 50 percent fat and 20 percent protein.

They are also rich in calcium and contain useful quantities of carotene, iron and B vitamins. Sesame seeds can form a nutritious addition to the diet.

Sunflower seeds

Sunflowers are grown mainly as a cash crop, but some of the seeds and some of the oil are eaten locally. The oil has the advantage of being relatively high in polyunsaturated fatty acids. The seeds contain about 36 percent oil (less than sesame), 23 percent protein and some calcium, iron, carotene and B vitamins.

Other oilseeds

A number of other oil-rich seeds are eaten or used for oil extraction. These include pumpkin seeds, melon seeds, oyster nut (Telfairia pedata) and cottonseed. The last is a major source of oil in the cotton-growing areas of Asia, Africa and Latin America. In West Africa and elsewhere, shea butter (Butyrospermum parkii),butternut and several other oilseeds are used in the diet. Most of these grow on indigenous trees.

Vegetables and Fruits

Vegetables

The foods called vegetables include some fruits (e.g. tomatoes and pumpkins), leaves (e.g. amaranth and cabbage), roots (e.g. carrots and turnips) and even stalks (e.g. celery) and flowers (e.g. cauliflower). Many of the plants from which these various edible parts are taken are unrelated botanically. However, "vegetable" is a useful term both in nutrition and in domestic terminology.

In developing countries, nearly all types of vegetables are eaten soon after they are harvested; unlike cereals, tubers, starchy roots, pulses and nuts, they are rarely stored for long periods (with a few exceptions such as pumpkins and other gourds).

It is not uncommon for rural people in parts of Asia, Latin America and Africa to forage for an important proportion of the vegetables they consume. With increasing population, however, the availability of wild fruits and vegetables is decreasing. Therefore vegetables are obtained from the farm or the household garden or from the marketplace, neighbours or small stalls along the roadside. When rural families with low income move to an urban environment they may resent having to purchase vegetables, because they are used to being able to gather wild ones or grow their own. They may therefore spend relatively little on this component of the diet. In any case, vegetables are rarely a prestige food, and in few societies are they high on the list of food preferences.

Vegetables are a very important part of the diet. They are nearly all rich in carotene and vitamin C and contain significant amounts of calcium, iron and other minerals. Their content of B vitamins is frequently small. They usually provide only a little energy and very little protein. A large proportion of their content consists of indigestible residue, which adds bulk or fibre to the faeces.

In many tropical diets the dark green leaves are the most valuable vegetables because they contain far more carotene and vitamin C, as well as more protein, calcium and iron, than pale green leaves and other vegetables. Thus amaranth is much superior to cabbage or lettuce. Leaves from pumpkin, sweet potato and cassava plants, as well as many wild edible leaves, are also excellent.

An increase in the consumption of green leaves and other vegetables could play a major part in reducing vitamin A deficiency, which is often prevalent in children, and could contribute to lessening the prevalence of iron deficiency anaemia in all segments of the population but especially in women of child-bearing age. Increased vegetable consumption would also supply additional calcium and vitamin C which would prevent the rare disease scurvy and perhaps also assist the healing of ulcers and wounds. Vitamin C also enhances iron absorption.

It is not possible here to describe the individual properties of the many vegetables commonly eaten in developing countries. A few, such as pumpkins, can be stored for several months with little loss of nutritive value; others, such as leaves and even tomatoes, are frequently sun-dried, but with considerable loss of vitamin content. The vitamin C content of vegetables is also lowered by prolonged cooking.

Vegetables grown in home and school gardens could be a valuable source of food for the family and the school and could make an important nutritional contribution, particularly to micronutrient intake. Home gardens can be raised with spare family labour and the participation of women and children. It is therefore important for most rural households and virtually every school to devote more time to growing vegetables. A community garden near the village source of water is often a useful adjunct to the villagers' own backyard gardens. In towns, even the smallest piece of land behind a house could, with the assistance of waste water, yield a valuable supply of vegetables all year round. The allocation of allotment plots for vegetable growing deserves serious consideration by town councils and other urban authorities. Even people living in flats can grow certain varieties in pots kept on their verandas.

Fruits

A wide variety of fruits grow wild or are cultivated in tropical countries. The varieties available at any one time in a given area depend on the climate, the local tastes for fruit, the species cultivated and the season.

The main nutritive value of fruits is their content of vitamin C, which is often high. Some fruits also contain useful quantities of carotene. Fruits (except the avocado and a few others) contain very little fat or protein and usually no starch. The carbohydrate is present in the form of various sugars. Fruits, like vegetables, contain much unabsorbable residue, mainly cellulose. The citrus fruits, such as oranges, lemons, grapefruits, tangerines and limes, contain good quantities of vitamin C but little carotene. In contrast, papayas, mangoes and Cape gooseberries (Physalis peruviana) contain both carotene and vitamin C.

Papayas are a useful fruit, especially for those who cultivate a piece of land for a few years and then move on to new land. The papaya grows rapidly and may yield fruit after one or two years. The mango, on the other hand, grows slowly, but once established (and it may establish itself) needs no care and yields fruit for half a century. Guavas, which are quite widely grown, contain five times as much vitamin C as most citrus fruits, as well as useful amounts of carotene.

The avocado requires special mention because, unlike other fruits, it is rich in fat, a substance that is lacking in many tropical diets. It could with benefit be much more widely grown and eaten and fed to children.

Bananas are widely grown and eaten in tropical countries. They contain fair quantities of carotene and vitamin C, and they are rich in potassium. In East Africa plantains or bananas are commonly picked when green. Cooked and eaten as a mainly starchy food, they form the staple diet of many people. When bananas are ripe their starch is converted into other sugars.

Foods of Animal Origin

Foods of animal origin are not essential for an adequate diet, but they are a useful complement to most diets, especially to those in developing countries that are based mainly on a carbohydrate-rich staple food such as a cereal or root crop. Meat, fish, eggs, milk and dairy products all provide protein of high biological value, which is often a good complement to the limiting amino acids in plant foods consumed. These products are also rich in other nutrients. The iron provided by meat and fish is easily absorbed and enhances the absorption of iron from common staple foods such as rice, wheat or maize. However, foods of animal origin are usually relatively expensive and not within the purchasing power of poorer families. Some wealthier people in both developing and industrialized countries consume large quantities of these foods; in consequence their intake of fat, especially saturated fat, may become excessive, increasing the risks of heart disease and obesity. Americans consume about 80 kg of meat per person per year - almost 0.25 kg per day.

Meat and Meat Products

Meat is usually defined as the flesh (mainly muscles) and organs (for example, liver and kidneys) of animals (mammals, reptiles and amphibians) and birds (particularly poultry). Meat is sometimes subdivided into red meat (from cattle, goats, sheep, pigs, etc.) and white meat (mainly from poultry). The animals providing meat may be domesticated or wild. The amount of meat consumed often depends mainly on cultural factors, on the price of meat in relation to incomes and on availability.

Meat contains about 19 percent protein of excellent quality and iron that is well absorbed. The amount of fat depends on the animal that the meat comes from and the cut. The energy value of meat rises with the fat content. The fat in meat is fairly high in its content of saturated fatty acids and cholesterol. Meat also provides useful amounts of riboflavin and niacin, a little thiamine and small quantities of iron, zinc and vitamins A and C. Offal (the internal organs), particularly liver, contains larger quantities. Offal has a relatively high amount of cholesterol. In general all animals - wild and domestic, large and small, birds, reptiles and mammals - provide meat of rather similar nutritional value. The main variable is the fat content.

Worldwide, a vast range and variety of animal products are eaten. Not all of them are popular everywhere, of course. Certain foods that are popular in some parts of the tropics and East Asia - such as locusts, grasshoppers, termites, flying ants, lake flies, caterpillars and other insects; baboons and monkeys; snakes and snails; rats and other rodents; and cats and dogs - are not found in European or North American diets. Similarly, the French liking for frogs' legs and horse meat and the English and Japanese taste for eels and raw oysters are not shared by many people living elsewhere. Liked or disliked, however, all these foods are nutritious and contain protein of high biological value.

Contaminated meat can lead to disease. There is a need for improvements in conditions associated with production of meat both for local or family consumption and more importantly for commercial sale. For meat to be safe for human consumption, hygienic practices are essential at all levels, from the farm, through the slaughterhouse, to the retailer and into the kitchen. Most countries have regulations governing meat hygiene and authorities responsible for applying the regulations, but their effectiveness varies widely.

Fish and Seafoods

Fish and seafoods, like meat, are valuable in the diet because they provide a good quantity (usually 17 percent or more) of protein of high biological value, particularly sulphur-containing amino acids. They are especially good as a complement to a cassava diet, which provides little protein.

Fish varies in fat content but generally has less fat than meat. Fish also provides thiamine, riboflavin, niacin, vitamin A, iron and calcium. It contains a small quantity of vitamin C if eaten fresh. Small fish from the sea and lakes such as sardines and sprats (dagaa in the United Republic of Tanzania, kapenta in Zambia) are consumed whole, bones and all, thus providing much calcium and fluorine. Dried dagaa, for instance, may contain 2 500 mg calcium per 100 g. Fish offal is not usually consumed as part of any diet anywhere. However, fish liver and fish oils are very rich sources of the fat-soluble vitamins A and D. The amount varies, usually with the age and species of fish.

Wherever water is available, fish provide a simple way of increasing protein consumption. The stocking of dams, the construction of fish ponds and better and more widespread fishing in rivers, lakes and the sea should all be given greater encouragement.

There is much regional variation in the variety of sea creatures people will eat. Encouraging children in coastal districts to collect sea urchins, sea slugs, limpets and the numerous other edible sea creatures, just as inland children collect locusts and lake flies, would considerably improve poor diets. The introduction of swimming lessons in youth clubs and as a community development activity would encourage development of this pastime as well as fishing both for pleasure and for profit; fear of the water because of inability to swim is a deterrent to these activities, particularly among people who do not live beside water.

Eggs

The egg is one of the few foods containing no carbohydrate. Just as the foetus in the mother's uterus draws nutrients from the mother's blood in order to grow and develop into a human being, so the bird embryo draws all its nutrients from within the egg. It is not surprising therefore that eggs are highly nutritious. Each egg contains a high proportion of excellent protein, is rich in fat and contains good quantities of calcium, iron, vitamins A and D and also thiamine and riboflavin.

Eggs are an essential part of the reproductive cycle of birds, so it is hardly surprising that their consumption, particularly by females, is forbidden by taboos in many societies. The irony is that eggs are often more easily available than most other high-quality foods. In developing countries it is not often that a family can afford to kill a cow or even a goat for food, but eggs are small and frequently laid. They are also an easily prepared, easily digestible, protein-rich food suitable for children from the age of six months onward. Eggs do have a nutritional disadvantage: very high cholesterol content. The cholesterol is present in the yolk.

Production of eggs for family use should be encouraged wherever possible, even in the small garden or yard of an urban dwelling. Toddlers should be given priority in eating the eggs.

Blood

Cattle blood, which is regularly consumed raw by many pastoral peoples, particularly in Africa, is highly nutritious. It is rich in protein, has high biological value and contains many other nutrients. It is a particularly valuable source of iron. It is also a good source of nutrients in its processed form, usually a type of sausage.

Milk and Milk Products

Animal milks and other dairy products are highly nutritious and can play an important part in human diets for both children and adults. The composition of milk varies according to the animal from which it comes, providing the correct rate of growth and development for the young of that species. Thus, for human infants, human milk is better than cows' milk or any other milk product. Exclusive breastfeeding without other foods or liquids is the optimum means of feeding for the first six months of an infant's life. Continuing breastfeeding for many more months is of great value, while the baby is introduced to other foods. If breastmilk remains an important food for the child into the second or even third year of life, then animal milk is not necessary in the child's diet.

Except for certain vitamins, the composition of human breastmilk is fairly constant, regardless of the diet of the mother. Maternal malnutrition will not cause a mother to produce milk of markedly lower nutrient content, but it will reduce the quantity she can produce. A few nutrients such as thiamine and vitamin A may be low if mothers are deficient in these nutrients.

Caseinogen and lactalbumin, proteins of high biological value, are among the most important constituents of cows' milk. The carbohydrate in cows' milk is the disaccharide lactose. Fat is present as very fine globules, which on standing tend to coalesce and rise to the surface. The fat has a rather high content of saturated fatty acids. The calcium content of cows' milk (120 mg per 100 ml) is four times that of human milk (30 mg per 100 ml), because calves grow much more quickly and have a larger skeleton than human babies and therefore need more calcium. When a human infant is fed entirely on cows' milk the excess calcium does no good but causes no harm. It does not produce a rate of growth beyond the optimum. The excess is excreted in the urine.

Milk is also a very good source of riboflavin and vitamin A. It is a fair source of thiamine and vitamin C, but it is a poor source of iron and niacin. The mother usually provides her infant with a store of iron before birth. However, this store is exhausted by about the sixth month of life, and if feeding of milk alone is prolonged, iron deficiency anaemia may develop.

The amount of thiamine in human milk varies more than the other constituents and is largely dependent on the mother's intake of this vitamin.

Infantile beriberi may occur in infants breastfed by thiamine-deficient mothers. The vitamin A content of human milk is to some extent dependent on the diet of the mother.

Despite the variation in the composition of milk from different animals, all milk is rich in protein and other nutrients and constitutes a good food for humans, especially children. Although most animal milk for human consumption comes from cows, in certain societies the milk of buffaloes, goats, sheep and camels is important. Some peoples have taboos against milk.

In many parts of the world, milk is more often consumed sour or curdled than fresh; in fact, some people dislike fresh milk. There is no need to alter this habit, for curdled milk keeps longer, retains its nutritive value and may be more digestible and more hygienic than fresh milk. However, it is much safer to drink milk that has been boiled and kept in a clean container, because milk can provide a vehicle for the transmission of some disease-causing organisms.

Pasteurization of milk carried out efficiently in a large, well-run dairy greatly reduces the risk of pathological organisms spreading, provided that the milk is placed in clean containers destined for direct delivery to the consumer. However, in many small towns where pasteurization is not well controlled, the milk may be insufficiently heated, the containers may not be well cleaned, and the milk may go from the plant into large churns for bottling elsewhere in insanitary surroundings. The consumer should not be overconfident in all milk labelled "pasteurized", since it is not necessarily free from pathological organisms.

In many countries where cows' milk is a normal item of the diet, it is customary to wean infants from breastmilk on to a diet in which cows' milk plays an important part. This is a valuable practice, for it helps ensure that the child will receive a balanced diet that provides all the requirements for growth, development and health.

Some people limit their milk consumption because of lactose intolerance, a condition resulting from low levels of the digestive enzyme lactase, which is responsible for digesting lactose, the main carbohydrate in milk. It is probably normal for human adults to have low levels of intestinal lactase, and the condition is very common in non-white peoples. Research shows that most lactose-intolerant persons can in fact consume milk in moderate quantities (perhaps three to five cups of milk per day) without developing symptoms.

Skimmed milk and dried skimmed milk

Skimmed milk is milk from which the fat has been removed, usually for making butter. In its dried form (DSM), it is a familiar product in many countries. It contains nearly all the protein of milk, as well as the carbohydrate, calcium and B vitamins. It is an excellent food, especially for those on predominantly carbohydrate diets and those who have extra needs for protein. In some places DSM is supplied to those with special needs through clinics and health centres. It is extensively used in hospitals and dispensaries as the basis for the treatment of protein-energy malnutrition (PEM). It is also issued at child-welfare clinics to prevent this most devastating form of malnutrition. Skimmed milk is an excellent food to add to any diet, but it is particularly useful in the diets of children and pregnant and lactating women. However, it is not a suitable substitute for whole milk for infants. It is sometimes added to dietary supplements such as, for example, corn (maize)/soybean/milk mixture (CSM).

Whole powdered milk

This product, as the name implies, is whole milk that has been dried. Unlike DSM, it contains fat. It is suitable for infants when no breastmilk is available.

Evaporated and condensed milks

These are milks that have had much of their water removed but that are still liquid. Condensed milk is sweetened by the addition of sugar, whereas evaporated milk does not contain added sugar. Many brands of condensed milk have vitamins added. These brands should be preferred to those that do not have vitamins added, especially if they are used in the diets of young children. They are not suitable as breastmilk substitutes for infants.

Yoghurt and soured or fermented milks

Many different organisms are used in the process of making yoghurt and fermented milks. These products are easy to prepare, are highly nutritious, have enhanced keeping quality and are a little less likely than fresh milk to harbour pathogenic organisms. Their use should be encouraged.

Casein

Casein is the protein from milk. It tends to be rather expensive. It is commonly mixed as part of a formula or mixture for treatment of children with PEM.

Cheese

The making of cheese no doubt arose from the desire of farm people to preserve some of the excess milk of the summer. Numerous processes are used, but essentially cheese is made by letting milk clot and subsequently removing some of the water. Salt and other flavourings may be added. Cheese-making is an excellent way of using any excess milk produced during the seasons when milk yields are high.

Oils and Fats

In general adults should consume at least 15 percent of their energy intake from dietary fats and oils, and women of childbearing age should consume at least 20 percent. Active individuals who are not obese may consume up to 35 percent and sedentary individuals up to 30 percent of energy from fat as long as saturated fatty acids do not exceed 10 percent of the energy intake and cholesterol intake is limited to 300 mg per day.

Infants fed human milk or formula usually receive 50 to 60 percent of their total energy from fat. Infants should receive breastmilk, but if they do not, the fatty acid composition of infant formula should correspond to the range found in the breastmilk from omnivorous women. During complementary feeding up to two years of age or beyond, the diet should provide 30 to 40 percent of energy from fat.

To achieve the recommended levels of fat intake, poor people, particularly in developing countries, would need to increase their intake of fat and oils. In contrast, most people living in rich industrialized countries would need to reduce their consumption of fat and oils, which now often provide 40 percent or more of the energy they consume.

The fat consumed in human diets is often divided into two categories: "visible" fat such as cooking oil and "invisible" fat such as the oil naturally present in cereals and legumes. Persons in developing countries who may get only 15 percent of their energy from fat will often obtain two-thirds as invisible fat and one-third as visible fat (or fat added to food). In contrast, in North America and Europe, where mean intakes of fats are high, some 70 percent may be visible fat and 30 percent invisible fat.

A diet very low in fat tends to be unpalatable and dull. It is difficult to cook a really good meal without any fat or oil, although the desired amount is largely a matter of habit and taste. However, like animal proteins, fats are relatively expensive, so the diet of poorer people is often short of fat. Fat is important because weight for weight it provides more than twice as much energy as carbohydrate or protein, thus reducing the bulk of the diet. Fats and oils may be good sources of fat-soluble vitamins, and they assist with the absorption of other nutrients. Recent work has established that certain unsaturated fatty acids are essential for pre- and postnatal development of the brain in children and are also essential for health in adults.

Fats contain a variety of fatty acids. Fats derived from land animals (e.g. butter and lard) usually contain a high proportion of saturated fatty acids and are solid at room temperature. Fats derived from vegetable products and marine animals (e.g. groundnut and cod-liver oils) contain more unsaturated fatty acids; they are usually liquid at room temperature and are termed oils. Coconut oil is an exception in that it contains mainly saturated fatty acids. A high intake of saturated fatty acids may contribute to raised serum cholesterol levels, which in turn may increase the risk of coronary heart disease.

Butter

Butter consists mainly of the fat from milk. It usually contains about 82 percent fat, with a trace of protein and carbohydrate; the rest is water. Butter is rich in vitamin A and has a small amount of vitamin D, but the content varies with the time of year and the diet of the cow from which it was derived. Usually about 800 mg of retinol and 50 IU of vitamin D are present in 100 g of butter. Butter and margarine are increasingly used in diets in developing countries as the use of bread increases.

Margarine

Developed as a substitute for butter, margarine is made from various vegetable oils that are partially hydrogenated to give a product with a consistency similar to that of butter. In most countries vitamins A and D are added so that the final product is nutritionally very similar to butter. If these vitamins have been added, they will usually be mentioned on the margarine container.

Ghee

Ghee is made by heating butter to precipitate the protein, which is then removed. Ghee contains 99 percent fat, no protein or carbohydrate, about 2 000 IU of vitamin A per 100 g and some vitamin D. It has good keeping qualities and is much used in tropical countries in place of butter, because butter soon goes rancid if kept unrefrigerated in warm temperatures.

Lard

Lard is collected during the heating of pork. Like other similar animal fats (e.g. drippings, suet), it consists of 99 percent fat and contains no carbohydrate, proteins, vitamins or minerals.

Vegetable Oils

Vegetable oils are the cooking fats most commonly used in Africa, Asia and Latin America, and there are many different kinds. Except for red palm oil, they have the disadvantage of containing no vitamins except vitamin E. They are mainly low in saturated fatty acids.

Commonly used vegetable oils are soybean, olive, maize, groundnut, sunflower, sesame, cottonseed and coconut oils. In their pure form, they are 100 percent fat and contain no water or other nutrients.

Red palm oil is widely produced in West Africa and in certain Asian countries (e.g. Malaysia). In West Africa it is important in human diets, but elsewhere it is exported for soap production and not much consumed locally. The oil contains large quantities of carotene, the precursor of vitamin A, commonly 12 000 μg per 100 g (with a range from 600 to 60 000 μg per 100 g). It is therefore a very valuable food wherever a shortage of vitamin A occurs in the diet. Vitamin A deficiency will not be a problem in areas where all members of the family consume even small quantities of red palm oil. Encouragement should be given to its wider cultivation and consumption.

Beverages and Condiments

Beverages

It is essential that the human body receive water, yet the human taste prefers that much of this water be obtained in the form of beverages. These include beer, wine, spirits, fruit juices, tea, coffee, cocoa, synthetic sweetened soft drinks and aerated waters. Some of these beverages contain small amounts of drugs such as caffeine (tea, coffee and some colas) or alcohol in varying amounts (beer, wine and spirits), and some are sources of minerals and vitamins.

In most countries there are traditional beverages of great variety. In Africa many of these are made from cereal grains that have been soaked and sprouted. These beverages may or may not be alcoholic, and some are useful sources of B vitamins. In other parts of the world local beverages may be made from honey or coconut or any number of local products.

In the industrialized countries aerated soft drinks, often called "sodas", many with a cola base, are highly popular and consumed in huge quantities. In many parts of Africa, Asia, Latin America and the Near East, manufactured soft drinks and sodas are replacing traditional beverages. Most of these sodas provide no significant nutrients other than carbohydrates.

In contrast, fruit juices, either purchased or home-made from fresh fruit, usually contain useful amounts of vitamin C, and some provide carotene. They are good beverages, especially for children.

It is not uncommon to find mothers giving their babies and children orange squash or fruit-flavoured sodas because they were told at the clinic to give them fruit juice. These manufactured beverages are no substitute for fruit juice and will do the child no good; they are simply a waste of money.

Certain vitamin-rich proprietary beverages have been designed for infants and children. Their vitamin content is nearly always clearly stated on the label. They need to be used with caution, however. They are not necessary if the child is getting fresh fruit and vegetables, and they are often a very expensive way of providing vitamin C to a child. The advertising promoting them is pervasive, however, and can persuade mothers that they are useful.

Another major group of beverages comprises those usually consumed hot. Tea, which was probably first drunk in China, is now the favourite beverage of many people in Africa, the Near East and Europe. Coffee originated in Africa but is now drunk most in the Americas, Europe and the Near East. The two main types are Arabian, Coffea arabica, and robusta, Coffea canephora. In all regions of the world tea, coffee and to a lesser extent cocoa are popular beverages. All three provide small amounts of caffeine, which is a mild stimulant. None have any great nutritional significance. Tannin and polyphenols in tea may reduce iron absorption.

For thousands of years people from all continents have produced beverages that contain ethyl alcohol. Usually certain yeasts are used to ferment a local carbohydrate-rich food (for example, cereals or root crops), but fruits, palm sap, honey and other raw ingredients are also used. In the industrialized countries beer (often made from barley), wine (made from grapes) and various spirits (drinks with a relatively high alcohol content made by distillation) are very widely consumed, and this practice has spread to many countries of the South. Alcohol produces a good feeling for many who drink it, but it also impairs the senses, and it can be addictive. It can be claimed that alcohol consumed in moderation provides a sense of well-being and may improve social interaction; but alcohol in excess is a serious cause of automobile and other accidents, and alcoholism is a highly prevalent and very damaging disease in all continents of the world.

Animals and primitive men and women obtained most of their fluids in the form of water; then for thousands of years other beverages became the favourite drink for humans; and now there is almost a craze to drink "natural" or "spring" waters, either aerated or still. Many consumers believe that these waters, coming from springs, lakes, rivers or wells, have near-magical qualities and great nutritive value. This idea is false. Bottled water may contain small amounts of minerals such as calcium, magnesium and fluoride, but so does tap-water from many municipal water supplies. A study comparing popular brands of bottled water showed that they were in no way superior to New York tap-water. They have only the advantage of being safe in areas where tap-water may be contaminated. However, for low-income people bottled waters are very expensive, and boiling local water renders it safe at a much lower cost.

Condiments

Salt consists mainly of sodium chloride. It is the only mineral salt that humans customarily consume in a chemically pure form. The body has a definite need for sodium and chlorine. The amount of sodium chloride in the body is regulated by the kidneys. In hot countries a person doing heavy work may lose 15 g of sodium chloride in body sweat in one day. Urinary excretion ranges from 1 to 30 g or more per day. Despite this loss, salt is not essential in the human diet unless sweating is profuse, because sufficient sodium and chlorine can be obtained from food alone. Nevertheless, nearly all people use salt, obtaining it by digging, making or buying it, however small the income. Certainly a salt-free diet is unpalatable. Adults usually consume about 10 g of salt a day, but there are enormous variations. A high intake of salt may contribute to the development of hypertension or high blood pressure in some individuals.

Other spices and flavourings are of less physiological or nutritive importance. In all countries, in all ages, people have added such items to their food to improve and vary its taste. In Africa, Asia and Latin America a variety of wild leaves are used, partly for flavour, partly as vegetables per se; hot chilies, both red and green, are frequently used; and pepper and curry powder are popular additions to the sauce or stew accompanying the staple food. Few of these flavourings have much nutritional importance, but all serve to make the food more pleasing to the taste. They therefore both increase the appetite and assist digestion by stimulating the secretion of saliva and intestinal juices. With the march of so-called civilization, many of the traditional and natural condiments and herbs are being replaced by proprietary sauces and flavourings. Some of these are artificial chemical agents (for example, monosodium glutamate) and some are based on traditional spices (garlic, cloves, ginger, etc.).

Food Processing and Fortification

Humans are unique in the animal kingdom in that they harvest, store and process food that they have grown. Almost all animals harvest food, and many animals store it for later consumption, but they do not grow it or process it. In their evolution from the apes humans learned to grow food for their own sustenance and then to develop many processes to preserve the food or to increase its desirable characteristics, sometimes thus decreasing or improving its nutritional value.

People seek to preserve food and to improve its quality using a variety of techniques such as drying, canning, pickling, adding chemical preservatives, refrigeration, freezing and irradiation. The main aim of these processes is to allow foods to remain in good edible condition, without serious deterioration, for longer than would be possible if these preservation methods were not used. The processes include cooking; adding substances to improve the taste or appearance of the food; taking measures to make the foods more nutritious, for instance adding micronutrients or germinating grains; and removing undesirable constituents, including toxins. Some food processing techniques have multiple effects. For example, milling of cereal grains may make them less nutritious, but it may also make them easier to cook and digest and less likely to deteriorate on storage.

Today food processing includes both traditional and some more industrial and modern techniques. Almost all aspects of food processing have some relevance to nutrition.

Research, teaching and extension regarding modern techniques of food processing are within the domain of food scientists rather than nutritionists. Food science is a very important subject which is advancing rapidly not only in academic institutions but also in the food industry, where large manufacturers often have advanced food science laboratories. Many books deal with food science, and some are included in the Bibliography.

Cooking

In ancient times and in traditional societies everywhere, cooking was and is the main food processing technique used. Humans learned to harness and make fire, and cooking their food became a way to improve the quality of their diets. Cooking techniques have changed much over the years in some societies and very little in others. Many people still cook over open fires and on traditional stoves, but in contrast now almost a majority of households in Western Europe and North America have a microwave oven in the kitchen, a relatively new invention. Similarly, industry uses both old and new cooking methods.

Cooking is practised by almost everybody, everywhere. Except for fruits and some vegetables, most groups of foods are generally cooked before being eaten. In many African and Asian countries even vegetables are seldom eaten uncooked, and there is little tradition of eating salads. The practice of cooking vegetables probably helps protect consumers from diseases

spread by faecal contamination including parasitic, bacterial and viral infections of the gastro-intestinal tract. Most tropical fruits are eaten raw, but the exposed peel is not consumed so they do not present the same risk of infections. Bananas, mangoes, papayas and citrus fruits, for example, are not dangerous because their peel is not eaten.

Cooking of food is a universal practice mainly because it improves the taste of food, makes inedible foods edible or makes foods more digestible. Cooking also kills organisms, including many disease-causing microorganisms in food. Cooking of high-starch foods including cereals (rice, wheat, maize, etc., which for most of humankind provide the bulk of the energy and even protein consumed) and also potatoes, yams and cassava makes these foods palatable and also more digestible. Cooking of some foods removes undesirable compounds such as antinutrients, for example trypsin inhibitors in soybeans and undesirable constituents in cassava.

There is more to cooking than merely roasting, baking, grilling, or boiling of foods as gathered or harvested. It usually also involves mixing of foods or perhaps more commonly adding food items to the main food being cooked, which may alter the nutritional value of the main food but is usually intended to make the food, dish or meal taste better. For example, fat is added in frying; salt, sugar, fruit and other products may be added to baked foods; and often the staple food such as potatoes may be cooked in a stew or soup with added onions, tomatoes and small quantities of meat. Cooking can be an art. It makes food tasty and desirable, and in most societies sharing a meal with family and friends is a pleasurable social occasion and is expected to do more than just fill the belly, assuage hunger and provide essential nutrients; it fuels feelings of mutuality and underpins the sense of community.

For all its good points, cooking can have some negative nutritional effects. Frying foods at very high temperatures can destroy some vitamins and can produce undesirable components such as carcinogens in the food. Smoking of food can also produce such substances. Boiling some items in water that is then discarded can remove water-soluble vitamins.

Germination of Grains

There is intense interest now in the use of traditional germinating methods to produce malted foods. For many years people in the United Republic of Tanzania and other countries have allowed sorghum, millet and other cereals to germinate by soaking the grains in water for some hours, then keeping them damp for two or three days, and finally drying them, often by spreading them in the sun. The dried cereal grains are then pounded using a traditional large pestle and mortar. The resulting flour is stored, and small amounts are used mainly for brewing local beer (pombe). The dried germinated flour, known as kimea, is also used to thin and sour traditional porridges made from maize for child feeding. The kimea thins the porridge because it produces the enzyme amylase, which breaks down the starch.

Preservation of Food

Physical Methods

Cooling or freezing greatly prolongs the time it takes for many foods to spoil or become inedible. In this sense it is a very important method of food preservation. Refrigerators are now very common appliances in the homes of better-off people in developing countries and are found in the majority of houses in industrialized countries. Freezers are also widely used.

Other methods used traditionally, but also in industry, are drying and smoking of foods. Removal of water prevents or reduces the ability of organisms to grow and multiply on or in many foods. Organisms thus inhibited include moulds and their toxic products, such as aflatoxin, as well as microorganisms that spoil the food and produce undesirable odours and taste. Dry cereals store better, and dried fish remains edible for relatively long periods. Some foods, such as milk, are dried in factories so that the preserved product can be easily marketed, transported and made available for consumption.

Chemical methods

Food may be kept edible longer by the use of substances termed chemical preservatives. The most widely used in the home are salt (sodium chloride) and sugar, which most homemakers would not consider to be chemical preservatives. Foods with high levels of salt or sugar are less attacked by organisms and so are preserved. Industry also uses salt and sugar to preserve food.

Over 100 years ago chemicals not usually available in the home (as salt and sugar are) were introduced as food preservatives. Some of these are not now used because of fears of toxicity; others are deemed safe and are widely used. International meetings have discussed safety issues, and most industrialized countries have regulations which list permitted food preservatives and concentrations that can be used. Among the widely used preservatives are sulphur dioxide and benzoates, which are mainly effective in controlling moulds and yeasts, respectively. Baked goods such as bread are often preserved using propionic acid, which inhibits the attack and growth of moulds and then prolongs the time before spoilage. Meats, particularly salted meats such as bacon and ham, are further preserved with sodium nitrite and sometimes sodium nitrate.

Sterilization

Both in the home and in the factory, foods of almost all kinds are preserved by a process termed canning, although some are actually put in jars or bottles. In general the foods (vegetables, fruits, meat products and others) are sterilized by heating them to kill all living organisms and are sealed in a can or bottle while still hot. Sometimes salt and sugar are used as part of the process. Home canning or bottling of foods of animal origin, particularly meat or fish of any kind, can be risky. Highly resistant bacteria such as Clostridium botulinum can survive, produce toxins and cause very serious disease.

Microbiological methods

Fermentation, which involves chemical breakdown of substances by microorganisms such as yeasts and bacteria, is used traditionally to preserve foods or to improve their palatability in many countries, as is the case with soy products in Indonesia. The process is also used commercially, for example in the manufacture of yoghurt or commercial alcoholic beverages.

Fermentation using yeasts and other organisms which act on the carbohydrate in the food produces alcohol. Humans almost all over the world, without food science classes, have discovered this mechanism and have found that alcohol consumption can be mood changing and pleasant. Thus with any carbohydrate they have, they use some to make alcoholic beverages. The carbohydrate may be a common cereal such as wheat, rice, barley or sorghum, or it may be honey, used to make mead in ancient Britain and modern Africa; coconut sap to make coconut wine in Oceania; or cassava or plantain to make strong drinks called waragi and koinage in Uganda.

Yeast also acts on sugars to produce carbon dioxide gas in food. This principle is used to make bread.

In some foods non-disease-causing organisms are encouraged to multiply to sour the food. Souring results when the microorganisms produce acid from the carbohydrate. Souring foods to some extent prevents pathogenic or harmful organisms from multiplying in the food, which keeps it safer and makes it last longer. Common soured foods are dairy products such as sour milk and yoghurt; fermented soy products such as tempeh; and fermented cereal porridges, consumed in much of sub-Saharan Africa. In some cases souring enhances the nutrient content of the food.

In many countries, including China, pickling is widely used to preserve vegetables and vegetable products.

Other methods

A purely industrial method of food preservation is irradiation. In this process the food is exposed to radiation, usually gamma rays, which kills microorganisms and fungal spores. The food is then sealed and is safe until opened. Irradiation can also be used to prevent or delay sprouting of certain cereals, legumes or other seeds and so to increase their shelf-life. Although irradiated foods are generally regarded as safe, there remains considerable debate about possible hazards of the irradiation process itself.

Fortification

Fortification is a form of food processing that is of special interest to nutritionists. When properly used it can be a strategy to control nutrient deficiencies. The terms "fortification" and "enrichment" are often used interchangeably. Fortification has been defined as the addition of one or more nutrients to a food to improve its quality for the people who consume it, usually with the goal of reducing or controlling a nutrient deficiency. This strategy may be applicable in nations or communities where there is a problem or a risk of a deficiency of the nutrient or nutrients concerned.

In some instances fortification may be the easiest, cheapest and best way to reduce a deficiency problem, but care needs to be taken to avoid its excessive promotion as a general panacea for controlling nutrient deficiencies. The pros and cons of fortification need to be weighed in each circumstance. Even so, as a strategy to control micronutrient deficiencies in developing countries fortification has often been underutilized, whereas in many industrialized countries it is generally overused. Ironically, it may add nutrients that are not generally lacking for consumers who are not at much risk of deficiency of those nutrients.

Outsiders should not rush into recommending fortification in a particular country. Local professionals need to be much involved in the planning, implementation and monitoring of a fortification programme. It is important to have a very clear picture of the local situation: nutrient deficiencies, food habits, food preparation practices, food processing facilities, marketing practices, etc. Food fortification is easier with one food, such as salt, and where there are very few manufacturers. Under other circumstances fortification is possible, could work and might have a major role in improving nutritional status and reducing the risk of deficiencies, even at the local level. In the past people have tried to find one ideal food to fortify with vitamin A or iron. Now it is recommended that countries consider fortifying several foods at the same time.

Two kinds of fortification that have been highly effective in many countries are the addition of iodine to salt (iodization) and the addition of fluorine to water (fluoridation). In the latter case the fluoride in a municipal water supply is augmented to provide levels that are considered optimal (i.e. one part per million) in order to reduce the incidence of dental caries or tooth decay.

In industrialized countries, and to some extent in developing countries, fortification is used to adjust the nutrient content of processed foods so that nutrient levels are close to those of the food before processing. For example, highly milled cereals such as wheat flour may have nutrients added to replace those lost during the refining process. An alternative would be to insist, or even legislate, that cereals not be highly refined.

Food fortification offers an important strategy to help control, in particular, the three main micronutrient deficiencies, namely deficiencies of iodine, vitamin A and iron. In developing countries the greatest priority should be given to fortification with these nutrients. With iodine, fortification alone, in the form of salt iodization, is often the only strategy used. With vitamin A and iron, fortification should be used in combination with, not to the exclusion of, other interventions. Particular care needs to be given to possible toxic problems with vitamin A, especially in women who are pregnant or planning to conceive. The advantages of fortification over some of the other strategies for controlling vitamin A and iron deficiencies are often relatively ignored and deserve more attention.

Other micronutrient deficiencies are of some importance in some countries, and fortification may be a good strategy to reduce the prevalence of deficiencies of, for example, niacin, thiamine, riboflavin, folate, vitamin C, zinc and calcium.

A somewhat different kind of fortification is the addition of macronutrients to enrich food. Enrichment could involve the addition of fat or oil to increase the energy density of a food; the addition of amino acids to cereal products to improve protein quality; or the addition of protein, sugar or oil (as well as micronutrients) to a formulated food, for example a manufactured weaning food, or to a food supplement such as corn (maize)/ soybean/milk (CSM) for emergency feeding.

Criteria or principles for fortification

The following are some of the conditions, considerations and principles relevant for those planning to fortify one or more foods to improve nutritional status. They apply mainly to fortification as a strategy to tackle micronutrient deficiencies.

— *Known nutrient deficiency in the population.* Dietary, clinical or biochemical data must show that a deficiency of the nutrient being considered exists to some degree in significant numbers of the population when they consume their usual diet, or that a risk exists.

— *Wide consumption of the food to be fortified among the at-risk population.* The food or foods to be fortified must be consumed by significant numbers of the population who have a deficiency of the nutrient being considered for fortification. If the deficiency disease occurs only in the very poor but they seldom purchase the food that is fortified, then it will do little good. Thus, for example, fortifying a relatively expensive manufactured weaning food with vitamin A may not help the poor children who have the highest prevalence of xerophthalmia if their parents cannot afford to purchase that food.

— *Suitability of the food and nutrient together.* Adding the nutrient to the food must not create any serious organoleptic problems. The items must mix well and this mixing must not cause an undesirable chemical reaction, any disagreeable taste, colour or odour changes or any other unacceptable characteristics.

— *Technical feasibility.* It must be technically feasible to add the nutrient to the food to satisfy the preceding condition.

— *Limited number of food manufacturers.* It is very helpful in a national or even a local fortification programme if there are relatively few manufacturers or processors of the food being considered. For example, if there are hundreds of salt producers, an iodization programme will face major problems. Similarly, if there are many millers, fortifying cereals will be difficult.

— *No substantial increase in price of the food.* It is important to consider the impact of fortification on the price of the food to be fortified. If adding a nutrient greatly increases the price of the food, consumption of the food may decline, particularly among poor people whose families are at special risk of the deficiency. If fortification does increase the price of the food, then it is possible to consider subsidizing the cost.

— *Range of consumption of the food.* Attention should be given to the usual range of consumption of the food being considered for fortification. If there is a very wide range between the smallest amount consumed, perhaps by 25 percent of the population, and the greatest amount consumed, perhaps by another 25 percent of the population, it may be difficult to decide the nutrient level for fortification. If large numbers of people at risk of the nutrient deficiency consume very small amounts of the food, then they may not benefit much from fortification. If significant numbers of individuals consume so much of the food that they may get toxic amounts of the nutrient, then the food may be unsuitable. In general there is a range of consumption of salt, and mean intakes may be 20 g per day, but practically no one consumes 200 g per day, every day. It is important to avoid a situation where people receive undesirable amounts of added nutrients, particularly in the case of fat-soluble vitamins or nutrients known to be toxic in large amounts.

— *Legislation.* When a government is moving in earnest to control a serious micronutrient deficiency using fortification, the appropriateness of legislation needs to be considered. Many industrialized countries have legislation to ensure that required minimum levels of B vitamins and sometimes also iron are present in wheat flour and some other cereal products. Many countries in the North and South have legislation to require that all salt sold is iodized, usually at a particular level. Fluoridation of water supplies to certain levels has been mandated legally, sometimes by municipalities and sometimes nationally.

— *Monitoring and control of fortification.* Monitoring to provide information on fortification of foods is useful. It is particularly important where fortification is legislated. In this case failure to fortify adequately could lead to prosecution and penalization of delinquent food manufacturers. Monitoring by governments is dependent on the availability of laboratory facilities and the trained personnel. Many countries lack adequate laboratory facilities to monitor salt iodization, and salt merchants are often aware that they can sell salt that is not iodized at all or at the level required by law. A good monitoring system needs to include testing, perhaps at sentinel sites all over the country. In the case of fluoridation, cities often monitor the fluoride content of their water. It is useful if a national laboratory also monitors the level of fluoride in municipal water provided to consumers.

Methods of fortification and suitable foods

The technology of fortification is a complex topic, discussed in many publications. Many different methods are used, and the choice of method depends on both the nutrient and the food.A frequently used technique in a flour or a fine-grained product involves adding a nutrient premix at a measured rate into the powdery food as it flows at some stage in the processing. Thorough mixing is necessary. This method is suitable for mills and large processing plants. For smaller processing facilities, or even at the village level, packages of premix are supplied. Instructions are given on the proportions to useand on methods to ensure a good mixture.

Difficulties have been encountered in fortifying rice because it is mainly eaten in a granular or whole-grain form. Therefore adding a powder, which is easy with wheat flour, is not possible with rice. At least two methods have been used. In one, rice grains are coated or impregnated with the nutrients to be used. In the second, artificial rice grains fortified with the desired nutrients are mixed with the rice. The artificial grains have to be well made so that they appear similar to ordinary rice grains. In the Philippines some decades ago it was reported that prior to cooking many housewives removed and threw away the fortified artificial rice grains because they had a yellowish colour from the added thiamine and riboflavin.

Some nutrients such as the B vitamins are relatively easy to add. Although vitamin A deficiency is of great importance, vitamin A is less easily used than the B vitamins in fortification programmes, in part because it is fat soluble and not water soluble. It is also likely to become oxidized. The most simple means is to add vitamin A to cooking oils and margarine, but food technology has overcome the difficulties, and many different foods have been successfully fortified with vitamin A in both industrialized and non-industrialized countries. For quite different reasons iron fortification of foods has presented serious challenges. Many different iron salts have been used. Often those that humans utilize best, such as ferrous sulphate, present the greatest difficulties and serious organoleptic problems.

References

Aguilera, Jose Miguel and David W. Stanley. 1999. *Microstructural Principles of Food Processing and Engineering*. Springer.

Carpenter, Ruth Ann; Finley, Carrie E. 2005.*Healthy Eating Every Day*. Human Kinetics.

Katz, Solomon. 2003. *The Encyclopedia of Food and Culture*, Scribner.

McGee, Harold. 2004. *On Food and Cooking: The Science and Lore of the Kitchen.* New York: Simon and Schuster.

Marion Nestle, 2007. *Food Politics: How the Food Industry Influences Nutrition and Health*, University Presses of California.

4

Malnutrition

Malnutrition is the condition that results from taking an unbalanced diet in which certain nutrients are lacking, in excess (too high an intake), or in the wrong proportions. A number of different nutrition disorders may arise, depending on which nutrients are under or overabundant in the diet. In most of the world, malnutrition is present in the form of undernutrition, which is caused by a diet lacking adequate calories and protein. While malnutrition is more common in developing countries, it is also present in industrialized countries. In wealthier nations it is more likely to be caused by unhealthy diets with excess energy, fats, and refined carbohydrates. A growing trend of obesity is now a major public health concern in lower socio-economic levels and in developing countries as well.

The World Health Organization cites malnutrition as the greatest single threat to the world's public health. Improving nutrition is widely regarded as the most effective form of aid. Nutrition-specific interventions, which address the immediate causes of undernutrition, have been proven to deliver among the best value for money of all development interventions. Emergency measures include providing deficient micronutrients through fortified sachet powders or directly through supplements. WHO, UNICEF, and the UN World Food Programme recommend community management of severe acute malnutrition with ready-to-use therapeutic foods, which have been shown to cause weight gain in emergency settings. The famine relief model increasingly used by aid groups calls for giving cash or cash vouchers to the hungry to pay local farmers instead of buying food from donor countries, often required by law, to prevent dumping from hurting local farmers.

Long term measures include fostering nutritionally dense agriculture by increasing yields, while making sure negative consequences affecting yields in the future are minimized. Recent efforts include aid to farmers. However, World Bank strictures restrict government subsidies for farmers, while the spread of fertilizer use may adversely affect ecosystems and human health and is hampered by various civil society groups.

Malnutrition has shown to be an important concern in women, children, and the elderly. Because of pregnancies and breastfeeding, women have additional nutrient requirements. Children can be at risk for malnutrition even before birth, as their nutrition levels are directly tied to the nutrition of their mothers. Breastfeeding can reduce rates of malnutrition and mortality in children, and educational programs for mothers could have a large impact on these rates. The elderly have a large risk of malnutrition because of unique complications such as changes in appetite and energy level, and chewing and swallowing problems. Adequate elderly care is essential for preventing malnutrition, especially when the elderly cannot care for themselves.

Malnutrition increases the risk of infection and infectious disease, and moderate malnutrition weakens every part of the immune system. For example, it is a major risk factor in the onset of active tuberculosis. Protein and energy malnutrition and deficiencies of specific micronutrients (including iron, zinc, and vitamins) increase susceptibility to infection. Malnutrition affects HIV transmission by increasing the risk of transmission from mother to child and also increasing replication of the virus. In communities or areas that lack access to safe drinking water, these additional health risks present a critical problem. Lower energy and impaired function of the brain also represent the downward spiral of malnutrition as victims are less able to perform the tasks they need to in order to acquire food, earn an income, or gain an education.

Protein-Energy Malnutrition

Protein-energy malnutrition (PEM) in young children is currently the most important nutritional problem in most countries in Asia, Latin America, the Near East and Africa. Energy deficiency is the major cause. No accurate figures exist on the world prevalence of PEM, but World Health Organization (WHO) estimates suggest that the prevalence of PEM in children under five years of age in developing countries has fallen progressively, from 42.6 percent in 1975 to 34.6 percent in 1995. However, in some regions this fall in percentage has not been as rapid as the rise in population; thus in some regions, such as Africa and South Asia, the number of malnourished children has in fact risen. In fact the number of underweight children worldwide has risen from 195 million in 1975 to an estimated 200 million at the end of 1994, which means that more than one-third of the world's under-five population is still malnourished.

Failure to grow adequately is the first and most important manifestation of PEM. It often results from consuming too little food, especially energy, and is frequently aggravated by infections. A child who manifests growth failure may be shorter in length or height or lighter in weight than expected for a child of his or her age, or may be thinner than expected for height.

The term protein-energy malnutrition entered the medical literature fairly recently, but the condition has been known for many years. In earlier literature it was called by other names, including protein-calorie malnutrition (PCM) and protein-energy deficiency.

The term PEM is used to describe a broad array of clinical conditions ranging from the mild to the serious. At one end of the spectrum, mild PEM manifests itself mainly as poor physical growth in children; at the other end of the spectrum, kwashiorkor (characterized by the presence of oedema) and nutritional marasmus (characterized by severe wasting) have high case fatality rates.

It has been known for centuries that grossly inadequate food intake during famine and food shortages leads to weight loss and wasting and eventually to death from starvation. However, it was not until the 1930s that Cicely Williams, working in Ghana, described in detail the condition she termed "kwashiorkor" (using the local Ga word meaning "the disease of the displaced child"). In the 1950s kwashiorkor began to get a great deal of attention. It was often described as the most important form of malnutrition, and it was believed to be caused mainly by protein deficiency. The solution seemed to be to make more protein-rich foods available to children at risk. This stress on kwashiorkor and on protein led to a relative neglect of nutritional marasmus and adequate food and energy intakes for children.

The current view is that most PEM is the result of inadequate intake or poor utilization of food and energy, not a deficiency of one nutrient and not usually simply a lack of dietary protein. It has also been increasingly realized that infections contribute importantly to PEM. Nutritional marasmus is now recognized to be often more prevalent than kwashiorkor. It is unknown why a given child may develop one syndrome as opposed to the other, and it is now seen that these two serious clinical forms of PEM constitute only the small tip of the iceberg. In most populations studied in poor countries, the point prevalence rate for kwashiorkor and nutritional marasmus combined is 1 to 5 percent, whereas 30 to 70 percent of children up to five years of age manifest what is now termed mild or moderate PEM, diagnosed mainly on the basis of anthropometric measurements.

Causes and Epidemiology

PEM, unlike the other important nutritional deficiency diseases, is a macronutrient deficiency, not a micronutrient deficiency. Although termed PEM, it is now generally accepted to stem in most cases from energy deficiency, often caused by insufficient food intake. Energy deficiency is more important and more common than protein deficiency. It is very often associated with infections and with micronutrient deficiencies. Inadequate care, for example infrequent feeding, may play a part.

The cause of PEM (and of some other deficiency diseases prevalent in developing countries) should not, however, be viewed simply in terms of inadequate intake of nutrients. For satisfactory nutrition, foods and the nutrients they contain must be available to the family in adequate quantity; the correct balance of foods and nutrients must be fed at the right intervals; the individual must have an appetite to consume the food; there must be proper digestion and absorption of the nutrients in the food; the metabolism of the person must be reasonably normal; and there should be no conditions that prevent body cells from utilizing the nutrients or that result in abnormal losses of nutrients. Factors that adversely influence any of these requisites can be causes of malnutrition, particularly PEM. The aetiology, therefore, can be complex. Certain factors that contribute to PEM, particularly in the young child, are related to the host, the agent (the diet) and the environment. The underlying causes could also be categorized as those related to the child's food security, health (including protection from infections and appropriate treatment of illness) and care, including maternal and family practices such as those related to frequency of feeding, breastfeeding and weaning.

Some examples of factors involved in the aetiology of PEM are:

— the young child's high needs for both energy and protein per kilogram relative to those of older family members;

— inappropriate weaning practices;
— inappropriate use of infant formula in place of breastfeeding for very young infants in poor families;
— staple diets that are often of low energy density (not infrequently bulky and unappetizing), low in protein and fat content and not fed frequently enough to children;
— inadequate or inappropriate child care because of, for example, time constraints for the mother or lack of knowledge regarding the importance of exclusive breastfeeding;
— inadequate availability of food for the family because of poverty, inequity or lack of sufficient arable land, and problems related to intrafamily food distribution;
— infections (viral, bacterial and parasitic) which may cause anorexia, reduce food intake, hinder nutrient absorption and utilization or result in nutrient losses;
— famine resulting from droughts, natural disasters, wars, civil disturbances, etc.

Prematurity or low birth weight may predispose the child to the development of nutritional marasmus. Failure of breastfeeding because of death of the mother, separation from the mother or lack of or insufficient breastmilk may be causes in poor societies where breastfeeding is often the only feasible way for mothers to feed their babies adequately. An underlying cause of PEM is any influence that prevents mothers from breastfeeding their newborn infants when they live in households where proper bottle-feeding may be difficult or hazardous. Therefore promotion of infant formula and insufficient support of breastfeeding by the medical profession and health services may be factors in the aetiology of marasmus. Prolonged exclusive breastfeeding without the introduction of other foods after six months of age may also contribute to growth faltering, PEM and eventually nutritional marasmus.

The view that kwashiorkor is the result of protein deficiency and nutritional marasmus the result of energy deficiency is an oversimplification, as the causes of both conditions are complex. Both endogenous and exogenous causes are likely to influence whether a child develops nutritional marasmus, kwashiorkor or the intermediate form known as marasmic kwashiorkor. In a child who consumes much less food than required for his or her energy needs, energy is mobilized from both body fat and muscle. Gluconeogenesis in the liver is enhanced, and there is loss of subcutaneous fat and wasting of muscles. It has been suggested that under these circumstances, especially when protein intake is very low relative to carbohydrate intake (with the situation perhaps aggravated by nitrogen losses from infections), various metabolic changes take place which contribute to the development of oedema. More sodium and more water are retained, and much of the water collects outside the cardiovascular system in the tissues, which results in pitting oedema. The actual role of infection has not been adequately explained, but certain infections cause major increases in urinary nitrogen, which derives from amino acids in muscle tissue.

There is not yet broad agreement on the actual cause of the oedema that is the hallmark of kwashiorkor. Most researchers agree that potassium deficiency and sodium retention are important in the pathogenesis of oedema. Some evidence supports the classical argument that oedematous malnutrition is a sign of inadequate protein intake. For example, oedema, fatty liver and a kwashiorkor-like condition can be induced in pigs and baboons on a protein-deficient diet. Epidemiological evidence also shows higher rates of kwashiorkor in Uganda, where the staple diet is plantain, which is very low in protein, than in neighbouring areas where the staple food is a cereal.

Recently two new theories have been advanced to explain the cause of kwashiorkor. The first is that kwashiorkor is due to aflatoxin poisoning. The second is that free radicals are important in the pathogenesis of kwashiorkor; it has been hypothesized that most of the clinical features of kwashiorkor could be caused by an excess free radical stress. This new, relatively untested theory also suggests, however, that kwashiorkor, even if produced by free radicals, is likely to occur only in children who have inadequate food intake and are subjected to infection. Thus even if this theory were to be proved correct, it would merely explain a mechanism for the pathogenesis of kwashiorkor. It would not change the fact that improving diet and reducing infection lead to significant reduction in both kwashiorkor and nutritional marasmus. Neither the aflatoxin nor the free radical theory has been proved experimentally, nor is there adequate convincing research to uphold the view of individual dysadaptation as the cause of severe PEM. Surprisingly, no studies have been able to give conclusive proof of either similarities or differences in dietary consumption between children who develop kwashiorkor with oedema and those who show clinical signs of nutritional marasmus without any oedema.

In severe PEM there is usually biochemical evidence, and often clinical evidence, of micronutrient deficiencies, which is not surprising in a child or adult who consumes a grossly inadequate diet. In both nutritional marasmus and kwashiorkor (and also in moderate PEM), clinical examinations or biochemical tests often give clear evidence of, for example, vitamin A deficiency, nutritional anaemia and/or zinc deficiency. However, there is little indication that any one micronutrient deficiency is the main cause of PEM or is by itself responsible for the oedema of kwashiorkor.

Irrespective of which theory of aetiology may be proved correct, improving the quantity of food consumed, taking steps to ensure that diets are nutritionally well balanced and controlling infection all help to prevent PEM.

Mild and Moderate PEM

The condition of PEM is often likened to an iceberg, of which 20 percent is visible above the water and about 80 percent submerged. The severe forms of PEM - kwashiorkor, nutritional marasmus and marasmic kwashiorkor - constitute the top, exposed part of the iceberg: they are relatively easy for a doctor or health worker to diagnose simply from their clinical manifestations, described below. On the other hand, children with moderate or mild malnutrition often do not have clear clinical manifestations of malnutrition; rather, they are shorter and/or thinner than would be expected for their age, and they may have deficits in psychological development and perhaps other signs not easy to detect. Mild and moderate PEM are diagnosed mainly on the basis of anthropometry, especially using measurements of weight and height and sometimes other measurements such as arm circumference or skinfold thickness.

As shown by the iceberg diagram (Figure 1), the prevalence of highly visible, serious PEM (kwashiorkor, marasmic kwashiorkor and nutritional marasmus) is usually between about 1 and 5 percent, except in famine areas. In contrast, moderate and mild malnutrition in many countries of sub-Saharan Africa and South Asia add up to 30 to 70 percent. In these areas often only 15 to 50 percent of young children between six months and 60 months of age do not have evidence of PEM. The diagram illustrates that both energy deficiency and protein deficiency play a part, but that energy deficiency is more important. It suggests that protein deficiency plays a greater part in kwashiorkor and energy deficiency in nutritional marasmus.

The percentage of children classified as having severe, moderate and mild PEM depends on how these terms are defined. The two severe forms of malnutrition, kwashiorkor and nutritional marasmus, have very different appearances and clinical features as described below. It is generally agreed that the hallmark of kwashiorkor is pitting oedema, and the overriding feature of nutritional marasmus is severe underweight. Children who have both oedema and severe underweight are diagnosed as having marasmic kwashiorkor.

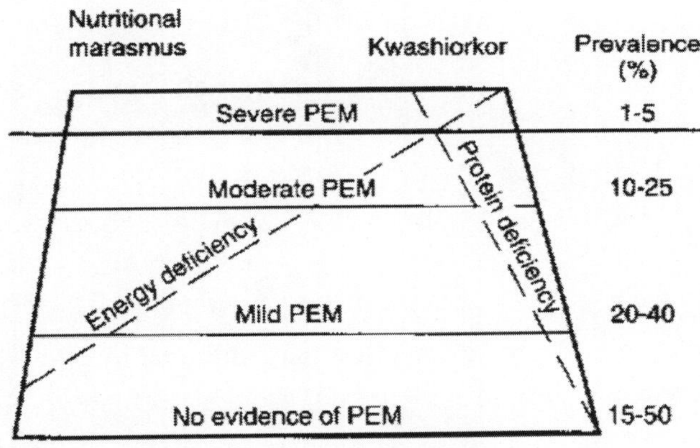

Figure 1. PEM iceberg

The so-called Wellcome classification of severe forms of PEM has been widely used for over 20 years. It has the advantage of simplicity because it is based on only two measures, namely the percentage of standard weight for age and the presence or absence of oedema. The category "undernourished" includes children who have moderate or moderately severe PEM but no oedema and whose weight is above 60 percent of the standard. Today a cut-off point using standard deviations (SD) is considered more appropriate than percentage of standard, but not many children would be reclassified.

In the 1950s and 1960s the degree of malnutrition was almost always based on the child's percentage of standard weight for age. In Latin America and elsewhere the Gomez classification was very widely used.In the early

1970s a number of nutrition workers began to suggest that judging the degree of malnutrition only on the basis of weight for age had many disadvantages. A method was suggested that distinguished three categories of mild to moderate PEM based on weight and height measurements of children. Subsequently these categories came to be known as follows:

— wasting: acute current, short-duration malnutrition, where weight for age and weight for height are low but height for age is normal;
— stunting: past chronic malnutrition, where weight for age and height for age are low but weight for height is normal;
— wasting and stunting: acute and chronic or current long-duration malnutrition, where weight for age, height for age and weight for height are all low.

This classification makes a distinction between current and past influences on nutritional status. It helps the examiner assess the likelihood that supplementary feeding will markedly improve the nutritional status of the child, and it gives the clinician some clue as to the history of the malnutrition in the patient. It also has advantages for nutritional surveys and surveillance. In general, stunting is more prevalent than wasting worldwide.

It is now generally recommended that malnutrition be judged on the basis of SD below the growth standards of the United States National Center for Health Statistics (NCHS) as published by WHO. In country reports published based on weight for age alone, "underweight" is commonly used to denote weight below 2 SD of the NCHS standards in children up to five years of age. In a normal distribution it is expected that 2 to 3 percent of children will fall below the -2 SD cut-off point. Prevalence above that level suggests that there is a nutritional problem in the population assessed. If measurements are also taken of length or height, then the children can be further divided into those who are wasted, stunted, or wasted and stunted.

Policy-makers and health workers need to decide which growth standards to use as a yardstick for judging malnutrition and for surveys, monitoring and surveillance. In recent years the WHO/NCHS growth standards (which do not differ very much from previously used standards, such as the Harvard and Denver growth standards) have gained increasing acceptance. The international growth standards have been found to be applicable for developing countries, as evidence shows that the growth of privileged children in developing countries does not differ importantly from these standards, and that the poorer growth seen among the underprivileged results from social factors, including the malnutrition-infection complex, rather than from ethnic or geographic differences.

The functional significance of mild or moderate PEM is still not fully known. Studies from several countries show that the risk of mortality increases rather steadily with worsening nutritional status as indicated by anthropometric measures. Recent investigations in Guatemala indicated that teenagers who had manifested poor growth when examined in early childhood were smaller in stature, did less well at school, had poorer physical fitness and had lower scores on psychological development tests than children from the same villages who grew better as young children. These results suggest long-term consequences of PEM in early childhood.

The attempt to control the extent and severity of PEM using many different strategies and actions is at the heart of nutritional programmes and policies in most developing countries. The reduction and eventual prevention of mild or moderate malnutrition will automatically reduce severe malnutrition. Thus, although it may be tempting (particularly for doctors and other health workers) to put major emphasis on the control of nutritional marasmus and kwashiorkor, resources are often better spent on controlling mild and moderate PEM, which will in turn reduce severe PEM.

Kwashiorkor

Kwashiorkor is one of the serious forms of PEM. It is seen most frequently in children one to three years of age, but it may occur at any age. It is found in children who have a diet that is usually insufficient in energy and protein and often in other nutrients. Often the food provided to the child is mainly carbohydrate; it may be very bulky, and it may not be provided very frequently.

Kwashiorkor is often associated with, or even precipitated by, infectious diseases. Diarrhoea, respiratory infections, measles, whooping cough, intestinal parasites and other infections are common underlying causes of PEM and may precipitate children into either kwashiorkor or nutritional marasmus. These infections often result in loss of appetite, which is important as a cause of serious PEM. Infections, especially those resulting in fever, lead to an increased loss of nitrogen from the body which can only be replaced by protein in the diet.

Nutritional Marasmus

In most countries marasmus, the other severe form of PEM, is now much more prevalent than kwashiorkor. In marasmus the main deficiency is one of food in general, and therefore also of energy. It may occur at any age, most commonly up to about three and a half years, but in contrast to kwashiorkor it is more common during the first year of life. Nutritional marasmus is in fact a form of starvation, and the possible underlying causes are numerous. For whatever reason, the child does not get adequate supplies of breastmilk or of any alternative food.

Perhaps the most important precipitating causes of marasmus are infectious and parasitic diseases of childhood. These include measles, whooping cough, diarrhoea, malaria and other parasitic diseases. Chronic infections such as tuberculosis may also lead to marasmus. Other common causes of marasmus are premature birth, mental deficiency and digestive upsets such as malabsorption or vomiting. A very common cause is early cessation of breastfeeding.

Treatment

Treatment of adult PEM includes therapy related to the underlying cause of the condition and therapy related to feeding and rehabilitation, when the cause makes that feasible. Thus infections such as tuberculosis or chronic amoebiasis require specific therapy which when effective will eliminate the cause of the weight loss and wasting. In contrast, curative treatment is not applicable in advanced AIDS or cancer.

Dietary treatment for adult PEM should be based on principles similar to those described for the treatment of severe PEM in children, including those recovering from kwashiorkor or marasmus. Emergency feeding and the rehabilitation of famine victims have relevance to adult PEM.

Prevention and Control of PEM

The prevention of PEM in Asia, Africa and the Americas presents a huge challenge. It is much more difficult than controlling, for example, iodine deficiency disorders (IDD) and vitamin A deficiency, because the underlying and basic causes, as described above, are often numerous and complex, and because there is no single, universal, cheap, sustainable strategy that can be applied everywhere to reduce the prevalence or severity of PEM.

In the late 1950s and 1960s it was thought that most PEM was caused mainly by inadequate intake of protein. A great deal of emphasis was placed on protein-rich foods as a major solution to the huge problem of malnutrition in the world. This inappropriate strategy diverted attention from the first need, which is adequate food intake by children. There is now much less emphasis on high-protein weaning foods and on nutrition education efforts to ensure greater consumption of meat, fish and eggs, which are economically out of the reach of many families who have children with PEM.

Protein is an essential nutrient, but PEM is more often associated with deficient food intake than with deficient protein intake. In general, when commonly consumed cereal-based diets meet energy needs, they usually also meet protein needs, especially if the diet also provides modest amounts of legumes and vegetables. Primary attention needs to be given to increasing total food intake and reducing infection.

Sensible efforts are needed to protect and promote breastfeeding and sound weaning; to increase the consumption by young children of cereals, legumes and other locally produced weaning foods; to prevent and control infection and parasitic disease; to increase meal frequency for children; and, where appropriate, to encourage higher consumption of oil, fat and other items that reduce bulk and increase the energy density of foods fed to children at risk. These measures are likely to have more impact if accompanied by growth monitoring, immunization, oral rehydration therapy for diarrhoea, early treatment of common diseases, regular deworming and attention to the underlying causes of PEM such as poverty and inequity. Some of these measures can be implemented as part of primary health care.

Nutritional Anaemias

Nutritional anaemias are extremely prevalent worldwide. Unlike protein-energy malnutrition (PEM), vitamin A deficiency and iodine deficiency disorders (IDD), these anaemias occur frequently in both developing and industrialized countries. The most common cause of anaemia is a deficiency of iron, although not necessarily a dietary deficiency of total iron intake. Deficiencies of folates (or folic acid), vitamin B_{12} and protein may also cause anaemia. Ascorbic acid, vitamin E, copper and pyridoxine are also needed for production of red blood cells (erythrocytes). Vitamin A deficiency is also associated with anaemia.

Anaemias can be classified in numerous ways, some based on the cause of the disease and others based on the appearance of the red blood cells. These classifications are fully discussed in medical textbooks.

Some anaemias do not have causes related to nutrition but are caused, for example, by congenital abnormalities or inherited characteristics; such anaemias, which include sickle cell disease, aplastic anaemias, thalassaemias and severe haemorrhage, are not covered here.

Based on the characteristics of the blood cells or other features, anaemias may be classified as microcytic (having small red blood cells), macrocytic (having large red blood cells), haemolytic (having many ruptured red blood cells) or hypochromic (having pale-coloured cells with less haemoglobin). Macrocytic anaemias are often caused by folate or vitamin B_{12} deficiencies.

In anaemia the blood has less haemoglobin than normal. Haemoglobin is the pigment in red cells that gives blood its red colour. It is made of protein with iron linked to it. Haemoglobin carries oxygen in the blood to all parts of the body. In anaemia either the amount of haemoglobin in each red cell is low (hypochromic anaemia) or there is a reduction in the total number of red cells in the body. The life of each red blood cell is about four months, and the red bone marrow is constantly manufacturing new cells for replacement. This process requires adequate amounts of nutrients, especially iron, other minerals, protein and vitamins, all of which originate in the food consumed.

Iron Deficiency

Iron deficiency is the most prevalent important nutritional problem of humans. It threatens over 60 percent of women and children in most non-industrialized countries, and more than half of these have overt anaemia. In most industrialized countries in North America, Europe and Asia, 12 to 18 percent of women are anaemic.

Although deficiency diseases are usually considered mainly as consequences of a lack of the nutrient in the diet, iron deficiency anaemia occurs frequently in people whose diets contain quantities of iron close to the recommended allowances. However, some forms of iron are absorbed better than others; certain items in the diet enhance or detract from iron absorption; and iron can be lost because of many conditions, an important one in many tropical countries being hookworm infection, which is very common.

Nutritional anaemias have until recently been relatively neglected and not infrequently remain undiagnosed. There are many reasons for the lack of attention, but the most important are probably that the symptoms and signs are much less obvious than in severe PEM, IDD or xerophthalmia, and that although anaemias do contribute to mortality rates they do not often do so in a dramatic way, and death is usually ascribed to another more conspicuous cause such as childbirth. However, research now indicates that iron deficiency has very important implications, including poorer learning ability and behavioural abnormalities in children, lower ability to work hard and poor appetite and growth.

Causes and epidemiology

To maintain good iron nutritional status each individual needs to have an adequate quantity of iron in the diet. The iron has to be in a form that permits a sufficient amount of it to be absorbed from the intestines. The absorption of iron may be enhanced or inhibited by other dietary substances.

Human beings have the ability both to store and to conserve iron, and it must also be transported properly within the body. The average male adult has 4 to 5 g of iron in his body, most of it in haemoglobin, a little in myoglobin and in enzymes and around 1 g in storage iron, mainly ferritin in the cells, especially in the liver and bone marrow. Losses of iron from the body must not deplete the supply to less than that needed for manufacture of new red blood cells.

To produce new cells the body needs adequate quantities and quality of protein, minerals and vitamins in the diet. Protein is needed both for the framework of the red blood cells and for the manufacture of the haemoglobin to go with it. Iron is essential for the manufacture of haemoglobin, and if a sufficient amount is not available, the cells produced will be smaller and each cell will contain less haemoglobin than normal. Copper and cobalt are other minerals necessary in small amounts. Folates and vitamin B_{12} are also necessary for the normal manufacture of red blood cells. If either is deficient, large abnormal red blood cells without adequate haemoglobin are produced. Ascorbic acid (vitamin C) also has a role in blood formation. Providing vitamin A during pregnancy has been shown to improve haemoglobin levels.

Of the dietary deficiency causes of nutritional anaemias, iron deficiency is clearly by far the most important. Good dietary sources of iron include foods of animal origin such as liver, red meat and blood products, all containing haem iron, and vegetable sources such as some pulses, dark green leafy vegetables and millet, all containing non-haem iron. However, the total quantity of iron in the diet is not the only factor that influences the likelihood of developing anaemia. The type of iron in the diet, the individual's requirements for iron, iron losses and other factors often are the determining factors.

Iron absorption is influenced by many factors. In general, humans absorb only about 10 percent of the iron in the food they consume. The adult male loses only about 0.5 to 1 mg of iron daily; his daily requirement for iron is therefore about 10 mg per day. On an average monthly basis, the adult pre-menopausal woman loses about twice as much iron as a man. Similarly, iron is lost during childbirth and lactation. Additional dietary iron is needed by pregnant women and growing children.

The availability of iron in foods varies greatly. In general, haem iron from foods of animal origin (meat, poultry and fish) is well absorbed, but the non-haem iron in vegetable products, including cereals such as wheat, maize and rice, is poorly absorbed. These differences may be modified when a mixture of foods is consumed. It is well known that phytates and phosphates, which are present in cereal grains, inhibit iron absorption. On the other hand, protein and ascorbic acid (vitamin C) enhance iron absorption. Recent research has shown that ascorbic acid mixed with table salt and added to cereals increases the absorption of intrinsic iron in the cereals two- to fourfold. The consumption of vitamin C-rich foods such as fresh fruits and vegetables with a meal may therefore promote iron absorption. Egg yolk impairs the absorption of iron, even though eggs are one of the better sources of dietary iron. Tea consumed with a meal may reduce the iron absorbed from the meal.

The normal child at birth has a high haemoglobin level (usually at least 18 g per 100 ml), but during the first few weeks many cells are haemolysed. The iron liberated is not lost but is stored in the body, especially in the liver and spleen. As milk is a poor source of iron, this reserve store is used during the early months of life to help increase the volume of blood, which is necessary as the baby grows. Premature infants have fewer red blood cells at birth than full-term infants, so they are much more prone to anaemia. In addition, iron deficiency in the mother may affect the infant's vital iron store and render the infant more vulnerable to anaemia. A baby's store of iron plus the small quantity of iron supplied in breastmilk suffice for perhaps six months, but then other iron-containing foods are needed in the diet. Although it is desirable that breastfeeding should continue well beyond six months, it is also necessary that other foods containing iron be introduced into the diet at this time.

Although most solid diets, both for children and adults, provide the recommended allowances for iron, the iron may be poorly absorbed. Many people have increased needs because of blood loss from hookworm or bilharzia infections, menstruation, childbirth or wounds. Women have increased needs during pregnancy, when iron is needed for the foetus, and during lactation, for the iron in breastmilk. It is stressed that iron from vegetable products, including cereal grains, is less well absorbed than that from most animal products.

Anaemia is common in premature infants; in young children over six months of age on a purely milk diet; in persons infected with certain parasites; and in those who get only marginal quantities of iron, mainly from vegetable foods. It is more common in women, especially pregnant and lactating women, than in men.

In most of the world, both North and South, the greatest attention to iron deficiency anaemia is directed at women during pregnancy, when they have increased needs for iron and often become anaemic. Pregnant women form the one group of the healthy population who are advised to take a medicinal dietary supplement, usually iron and folic acid. Pregnant and lactating women are a group at especially high risk of developing anaemia.

It is only in recent years that the prevalence and importance of iron deficiency apart from anaemia has been widely discussed. Clearly, however, if the causes of iron deficiency are not removed, corrected or alleviated then the deficiency will lead to anaemia, and gradually the anaemia will become more serious. Increasing evidence suggests that iron deficiency as manifested by low body iron stores, even in the absence of overt anaemia, is associated with poorer learning and decreased cognitive development.

International agencies now claim that iron deficiency anaemia is the most common nutritional disorder in the world, affecting over 1 000 million people. In females of child-bearing age in poor countries prevalence rates range from 64 percent in South Asia to 23 percent in South America, with an overall mean of 42 percent. Prevalence rates are usually considerably higher in pregnant women, with an overall mean of 51 percent. Thus half the pregnant women in these regions, whose inhabitants represent 75 percent of the world's population, have anaemia. Unlike reported figures for PEM and vitamin A deficiency, which are declining, estimates suggest that anaemia prevalence rates are increasing.

In most of the developing regions, and particularly among persons with anaemia or at risk of iron deficiency, much of the iron consumed is non-haem iron from staple foods (rice, wheat, maize, root crops or tubers). In many countries the proportion of dietary iron coming from legumes and vegetables has declined, and rather small quantities of meat, fish and other good sources of haem iron are consumed. In some of the regions with the highest prevalence of anaemia the poor are not improving their dietary intake of iron, and in some areas the per caput supply of dietary iron may even be decreasing year by year.

In many parts of the world where iron deficiency anaemia is prevalent it is due as much to iron losses as to poor iron intakes. Whenever blood is lost from the body, iron is also lost. Thus iron is lost in menstruation and childbirth and also when pathological conditions are present such as bleeding peptic ulcers, wounds and a variety of abnormalities involving blood loss from the intestinal or urinary tract, the skin or various mucous membrane surfaces. Undoubtedly one of the most prevalent and important causes of blood loss is hookworms, which can be present in very large numbers. The worms suck blood and also damage the intestinal wall, causing blood leakage. Some 800 million people in the world are infested with hookworms. Other intestinal parasites such as Trichuris trichiura may also contribute to anaemia. Schistosomes or bilharzias, which are of several kinds, also cause blood loss either into the genito-urinary tract (in the case of Schistosoma haematobium) or into the gut. Malaria, another very important parasitic infection, causes destruction of red blood cells that are parasitized, which can lead to what is termed haemolytic anaemia rather than to iron deficiency anaemia. In programmes to reduce anaemia actions may be needed to control parasitic infections and to reduce blood loss resulting from disease as well as to improve dietary intakes of iron.

Anaemia resulting from folate deficiency is less prevalent than that from iron deficiency or iron loss. It occurs when folate intakes are low and when red cells are haemolysed or destroyed in conditions like malaria. The anaemia of both folate and vitamin B_{12} is macrocytic, with larger than normal red blood cells. Folic acid or folates are present in many foods including foods of animal origin (e.g. Iiver and fish) and of vegetable origin (e.g. Ieafy vegetables). Vitamin B_{12} is present only in foods of animal origin. In most countries vitamin B_{12} deficiency is uncommon.

Clinical manifestations

Haemoglobin in the red blood cells is necessary to carry oxygen, and many of the symptoms and signs of anaemia result from the reduced capacity of the blood to transport oxygen. The symptoms and signs are:

— tiredness, fatigue and lassitude;
— breathlessness following even moderate exertion;
— dizziness and/or headaches;
— palpitations, with the person complaining of being aware of his or her heartbeat;
— pallor of the mucous membranes and beneath the nails;
— oedema (in chronic, severe cases).

These symptoms and signs are not confined to iron deficiency anaemia but are similar in most forms of anaemia. Most occur also in some other illnesses and thus are not specific to anaemia. Because none of the symptoms seem severe, dramatic or life threatening, at least in the early stages of anaemia, the disorder tends to be neglected.

An experienced health worker can sometimes make a preliminary diagnosis by examining the tongue, the conjunctiva of the lower eyelid and the nailbed, which may all appear paler than normal in anaemia. The examiner can compare the redness or pinkness below the nail of the patient with the colour beneath his or her own nails. Enlargement of the heart may result and can be detected in advanced severe anaemia. Oedema usually occurs first in the feet and at the ankles. There may also be an increased pulse rate or tachycardia. Occasionally the nails become relatively concave rather than convex and become brittle. This condition is termed koilonychia. Anaemia is also reported to lead both to abnormalities of the mouth such as glossitis and to pica (abnormal consumption of earth, clay or other substances).

What is surprising is that many persons with very low haemoglobin levels, especially women in developing countries, appear to function normally. With chronic anaemia they have adapted to low haemoglobin levels. They may indeed do reduced work, have fatigue and walk more slowly, but they still give the appearance of performing their normal duties even though severely anaemic. Severe anaemia can progress to heart failure and death.

Anaemia, as well as producing the symptoms and signs discussed above, also leads to a reduced ability to do heavy work for long periods; to slower learning and more difficulty in concentration by children in school or elsewhere; and to poorer psychological development.

A very important aspect of anaemia in women is that it markedly increases the risk of death of the mother during or after childbirth. The woman may bleed severely, and she has low haemoglobin reserves. There is also an increased risk for her infant.

Iodine Deficiency Disorders

Iodine deficiency is responsible not only for very widespread endemic goitre and cretinism, but also for retarded physical growth and intellectual development and a variety of other conditions. These conditions together are now termed iodine deficiency disorders (IDD). They are particularly important because:

— perhaps one-quarter of the world's people consume inadequate amounts of iodine;

— the disorders have a major impact on the individual and on society;

— of the four major deficiency diseases, IDD is the easiest to control.

In fact, as H.R. Labouisse wrote in 1978 when he was Executive Director of the United Nations Children's Fund (UNICEF), "Iodine deficiency is so easy to prevent that it is a crime to let a single child be born mentally handicapped for this reason". Nonetheless this crime persists.

Endemic goitre and severe cretinism are the exposed part of the IDD iceberg. These are abnormalities that are visible to the populations where they are prevalent, and they can be diagnosed relatively easily by health professionals without the use of laboratory or other tests. The submerged and larger part of the iceberg includes smaller, less visible enlargements of the thyroid gland and an array of other abnormalities. In many areas of Latin

America, Asia and Africa iodine deficiency is a cause of mental retardation and of children's failure to develop psychologically to their full potential. It is also associated with higher rates of foetus loss (including spontaneous abortions and stillbirths), deaf-mutism, certain birth defects and neurological abnormalities.

For several decades the main measure used to control IDD has been the iodization of salt, and when properly conducted and monitored it has proved extremely effective in many countries. It is also relatively cheap. Several international meetings, including the International Conference on Nutrition held in Rome in 1992, called for the virtual elimination of IDD by the year 2000. This goal is achievable, provided the effort receives international support and real national commitment in each of the many countries where the disorders remain prevalent.

Causes

The most important cause of endemic goitre and cretinism is dietary deficiency of iodine. The amount of iodine present in the soil varies from place to place and this influences the quantity of iodine present in the foods grown in different places and in the water. Iodine is leached out of the soil and flows into streams and rivers which often end in the ocean. Many areas where endemic goitre is or has been highly prevalent are plateau or mountain areas or inland plains far from the sea. These areas include the Alps, the Himalayas and the Rocky Mountains; smaller mountain ranges or highland areas in countries such as China, the United Republic of Tanzania, New Zealand, Papua New Guinea and countries of Central Africa; and inland areas and plains in the United States, Central Asia and Australia.A less important cause of IDD is the consumption of certain foods which are said to be goitrogenic or to contain goitrogens. Goitrogens are "antinutrients" which adversely influence proper absorption and utilization of iodine or exhibit antithyroid activity. Foods from the genus Brassica such as cabbage, kale and rape and mustard seeds contain goitrogens, as do certain root crops such as cassava and turnips. Unlike goitrogenic vegetables, cassava is a staple food in some areas, and in certain parts of Africa, for example Zaire, cassava consumption has been implicated as an important cause of goitre.

Epidemiology

Any enlargement of the thyroid gland is called a goitre. The thyroid is an endocrine gland centrally situated in the lower front part of the neck. It consists of two lobes joined by an isthmus. In an adult each lobe of the normal thyroid gland is about the size of a large kidney bean. In areas of the world or communities where only sporadic goitre occurs or where health workers see only an occasional patient with an enlarged thyroid gland, the cause is not likely to be related to the individual's diet. Sporadic goitre may for example be due to a thyroid tumour or thyroid cancer. However, if goitre is common or endemic in a community or district, then the cause is usually nutritional. Endemic goitre is almost certainly caused by iodine deficiency, and where goitre is endemic other iodine deficiency disorders can also be expected to be prevalent.

Where goitre is endemic, often large numbers of people have an enlargement of the thyroid gland, and some have enormous unsightly swellings of the neck. The condition is usually somewhat more prevalent in females, especially at puberty and during pregnancy, than in males. The enlarged gland may be smooth (colloid goitre) or lumpy (adenomatous or nodular goitre).

The iodine content of foods varies widely, but the amount of iodine present in common staple foods such as cereals or root crops depends more on the iodine content of the soil where the crop is grown than on the food itself. Because the amount of iodine in foods such as rice, maize, wheat or legumes depends on where they are grown, food composition tables cannot provide good figures for their iodine content. Foods from the ocean, including shellfish, fish and plant products such as seaweed, are generally rich in iodine.

In many populations, particularly in the industrialized countries of the North and among affluent groups almost everywhere, diets do not depend mainly on locally grown foods. As a result many of the foods purchased and consumed may contribute substantially to iodine intakes. For example, persons living in the Rocky Mountains of North America, where goitre used to be endemic, now do not rely much on locally produced foods; they may consume bread made from wheat grown in the North American central plains, rice from Thailand, vegetables from Mexico or California, seafood from the Atlantic coast and so on. Similarly, affluent segments of society in La Paz, Bolivia consume many foods not grown in the altiplano, and these imported foods will have adequate quantities of iodine. In contrast, the poor in the Bolivian highlands eat mainly locally grown foods and do develop goitre.

Many countries of Asia, Africa and Latin America have major iodine deficiency problems, although some countries have made great progress in reducing the prevalence of IDD. China and India, with their vast populations, still have a high prevalence of IDD. Not all African countries have been surveyed, but it is known that IDD is prevalent in Ethiopia, Nigeria, Tanzania, Zaire, Zimbabwe and many smaller nations. In the Americas, endemic goitre has been largely controlled in the United States and Canada, but many Andean countries including Bolivia, Colombia, Ecuador and Peru still have relatively high endemic goitre and cretinism rates. IDD is also encountered in the Central American countries and in parts of Brazil.

During a survey conducted by the author in the 1960s in the Ukinga Highlands of Tanzania, 75 percent of the people examined had goitre. This was the highest prevalence yet reported in Africa. Prevalence rates of over 60 percent have been reported from communities in many African, Asian and Latin American countries.

Generally goitre prevalence rates of 5 to 19.9 percent are considered mild, 20 to 29.9 percent moderate and 30 percent and over severe. But even with rates of 10 to 15 percent the need for action is important. Where prevalence rates are moderate, urgent action is needed. Where rates are severe, early action is critical.

Treatment

The treatment of goitre caused by iodine deficiency is easy and satisfying in the case of a simple goitre or a colloid goitre that is not very large. Usually either potassium iodide (6 mg daily) or Lugol's iodine (one drop daily for ten days, then one drop weekly) will lead to a fairly rapid reduction in the size of the goitre. One drop of Lugol's iodine provides about 6 mg of iodine. Alternatively, Lugol's iodine can be diluted in any small hospital laboratory so that one teaspoonful of the dilute solution yields 1 mg of iodine. Lugol's solution is very cheap and is widely available. Of primaryschool children treated in Tanzania, over 60 percent with Grade 1 goitre had no goitre after 12 weeks of receiving Lugol's iodine, and most larger goitres had improved markedly. An alternative treatment which is also effective but which needs careful medical supervision is the use of thyroid extract or medicinal thyroxine.

Large nodular goitres and some other goitres that do not respond to treatment with either iodine or thyroxine can only be properly treated by surgical excision. Surgery is especially needed if the goitre is causing symptoms because it is retrosternal or pressing on the trachea. Thyroidectomy requires a good well-trained surgeon and good medical management afterwards. Patients who have had total thyroidectomy must receive thyroxine or thyroid hormones for the rest of their lives.

Prevention of IDD

Clearly, rather than treating each individual who has goitre caused by iodine deficiency, it is much preferable to take measures to control iodine deficiency in the community, the district or the nation. The most common, and often the best, measure is iodization of salt, which will reduce the prevalence and also the severity of goitre over a relatively short period among those who consume the salt.

Vitamin A deficiency

Vitamin A was discovered in 1913 when experiments showed that if the only fat present in diets of young animals was lard, their growth was retarded, and when butter was substituted the animals grew and thrived. A substance in butter but not in lard was found also in egg yolk and cod-liver oil. It was named vitamin A. It was later established that many products of vegetable origin had nutritional properties similar to those presented by vitamin A in foods of animal origin; they were found to contain a yellow pigment, carotene, which is converted to vitamin A in the body. Preformed vitamin A or retinol is a fat-soluble vitamin found only in animal products. Carotenes or carotenoids can act as a provitamin. There are many carotenoids in plants, but the most important for human nutrition is beta-carotene, which can be converted to vitamin A by enzymatic action in the intestinal wall. Breastmilk is an important source of vitamin A for infants.

Dietary deficiency of vitamin A most commonly and importantly affects the eyes, and it can lead to blindness. Xerophthalmia, meaning drying of the eyes (from the Greek word xeros, meaning dry), is the term now used to cover the eye manifestations resulting from vitamin A deficiency. Vitamin A deficiency also has a role in a variety of clinical conditions not related to the eyes, and it may contribute to higher child mortality rates, especially in children who develop measles. It has been demonstrated that laboratory animals on diets deficient in vitamin A have increased rates and severity of infections. Vitamin A deficiency also adversely affects epithelial surfaces apart from the eye and is associated with an increased incidence of certain cancers, including cancer of the colon. The serious eye manifestations of vitamin A deficiency leading to corneal destruction and blindness are mainly seen in young children. This condition is sometimes called keratomalacia.

Until recently vitamin A deficiency was a relatively neglected condition, probably for the following four reasons:

- Public health and nutrition efforts were concentrated on the control of protein-energy malnutrition (PEM), with which vitamin A deficiency is associated, and which is the most important form of malnutrition in non-industrialized countries.
- Where xerophthalmia is prevalent there were few eye specialists or health workers who could correctly diagnose the condition.
- The condition occurs in the very young child behind closed eyelids, or it does not appear to the parents to warrant medical attention until too late, when the cornea is irreversibly damaged.
- Because the fatality rates from advanced xerophthalmia are high, relatively few blind children survive in the community, which reduces the social significance and visibility of the problem.

However, recently the World Summit for Children (1991) and the International Conference on Nutrition (1992) called for the virtual elimination of vitamin A deficiency and its consequences, including blindness, by the year 2000. Much more emphasis is now being placed on the control of vitamin A deficiency.

Causes

An inadequate intake of carotene or preformed vitamin A, poor absorption of the vitamin or an increased metabolic demand can all lead to vitamin A deficiency. Of these three, dietary deficiency is by far the most common cause of xerophthalmia.

Good sources of retinol, or preformed vitamin A, are liver, fish-liver oils, egg yolks and dairy products. In most non-industrialized countries, however, the majority of poor people get most, often 80 percent or more, of their vitamin A from carotene in foods of vegetable origin. The yellow colour of carotene may be masked by chlorophyll in many dark green leafy vegetables. Carotenes are present in good quantities in a wide variety of green and yellow vegetables and fruits, in yellow maize and in yellow root crops, e.g. sweet potatoes. A rich source is red palm oil, which is eaten extensively in West Africa and widely grown but infrequently consumed in many other areas, e.g. Malaysia. In many tropical diets important sources are dark green leafy vegetables [e.g. amaranth, cassava and drumstick (Moringa oleifera)leaves], mangoes, papayas, tomatoes and sometimes local yellow pumpkins, squash and yellow maize. The wet tropics often abound in both cultivated and wild food sources of carotene, but the poor often consume too little of these foods, and young children often dislike green vegetables. In some seasons the main sources of vitamin A may be less available or more expensive.

The biological activity of vitamin A is now usually expressed as retinol equivalents (RE) rather than in international units (IU). One RE is equal to 1 μg of retinol or 6 μg of beta-carotene. The World Health Organization (WHO) has recommended an intake of 300 RE daily for infants and 750 RE for adults.

Vitamin A, either preformed (retinol) or converted from carotene, is stored in the liver. Retinol is transported from the liver to other sites in the body by retinol binding protein (RBP), a specific carrier protein. Protein deficiency may influence vitamin A status by reducing the synthesis of RBP.

Low intake of vitamin A and carotene over an extended period is the most common cause of xerophthalmia. The condition may be influenced by other factors, however, e.g. intestinal parasitic infections, gastro-enteritis or malabsorption. Measles often precipitates xerophthalmia because it leads to lowered food intake (in which anorexia and stomatitis may be factors) and to increased metabolic demands for vitamin A. The virus may also affect the eye, aggravating lesions caused by vitamin A deficiency. PEM is also important as a cause or accompaniment of xerophthalmia. Data from Indonesia and elsewhere suggest that serious corneal involvement in xerophthalmia seldom occurs except in children who have moderate or severe PEM.

Epidemiology

Vitamin A deficiency is the most common cause of blindness in children in many endemic areas. Xerophthalmia occurs almost entirely in children living in poverty. It is extremely rare to find cases in more affluent families, even in areas where xerophthalmia is prevalent. It is a disease related to low socio-economic status, low levels of female literacy, land shortages, inequity, poor availability of curative and preventive primary health care, high rates of infectious and parasitic diseases (often related to poor sanitation and water supplies) and grossly inadequate family food security. As with PEM, three essentials for prevention of vitamin A deficiency are adequate food security, care and health.

It is always frustrating and extremely sad to see a child with advanced xerophthalmia that includes a perforated cornea when a few days earlier the sight of the child could easily have been saved. A few days and a few cents could have prevented a whole lifetime of blindness. The parents are often poor and uneducated. They love their children, but they may be resigned about the illness because they have inadequate access to good health care, and they may be fatalistic or suspicious of Western medicine. Therefore a small eye problem may not lead the parents to seek early health care even if it is easily available.

In recent decades xerophthalmia has been especially prevalent in children of poor rice-eating families in South and Southeast Asia (e.g. Bangladesh, India, Indonesia and the Philippines). There is a high incidence in some African countries (e.g. Burkina Faso, Ethiopia, Malawi, Mozambique and Zambia), whereas other countries, especially in West Africa, seem to have a lower prevalence in part because of the consumption of red palm oil, which is high in carotene. In the Western Hemisphere, Haiti and northeastern Brazil are areas where xerophthalmia is highly prevalent. It occurs also in many poorer areas of Central and South America. Vitamin A deficiency used to be a problem in the Near East, but few recent data on its prevalence there are available. In poor developing countries where vitamin A deficiency is endemic, it is also prevalent among lactating mothers. In Europe and North America, and in affluent people everywhere, vitamin A deficiency may occur in alcoholics, in those with malabsorption or anorexia nervosa and in persons who for any reason consume diets low in carotene or vitamin A.

Prevalence rates of five different signs have been recommended as criteria for judging whether xerophthalmia is a significant public health problem in a given populatio. It is suggested that if the prevalence of any one sign (i.e. the percentage of children examined having the sign) in children aged six months to six years in a vulnerable population is above the cut-off, then xerophthalmia should be considered a public health problem in that population.

It is believed that worldwide between 500 000 and 1 million children each year develop active xerophthalmia with some corneal involvement. Of these, perhaps half will become blind or have serious visual impairment, and a large proportion will die. In addition, many millions of children are vitamin A deficient or at risk but do not have xerophthalmic eye manifestations. Deficiency is manifested by low liver stores of retinol and low serum vitamin A levels.

Treatment

Effective treatment depends on early diagnosis, immediate dosing with vitamin A and proper treatment of other illnesses such as PEM, tuberculosis, infections and dehydration. Severe cases with corneal involvement should be treated as emergencies. Sometimes hours, and certainly days, may make the difference between reasonable vision and total blindness.

Treatment for children one year of age or over should consist of 110 mg of retinyl palmitate or 66 mg of retinyl acetate (200 000 IU of vitamin A) orally or preferably 33 mg (100 000 IU) of water-miscible vitamin A (retinyl palmitate) by intramuscular injection. Vitamin A in oil should not be used for injection. The oral dose should be repeated on the second day and again on discharge from hospital or seven to 30 days after the first dose. These doses should be halved for infants.

When there is corneal involvement it is desirable to apply an antibiotic ointment such as topical bacitracin to both eyes six times per day. Appropriate systemic antibiotics should also be administered.

Night blindness and conjunctival xerosis are completely reversible and respond quickly to treatment using oral doses of vitamin A on an out-patient basis. Corneal ulceration is arrested by treatment and will heal within a week or two but will leave scars. The case fatality rate is often high because of accompanying PEM and infections.

Prevention

In the long term, sustainable control will be achieved by increasing the production and consumption of foods rich in vitamin A and carotene by at-risk populations. Other methods include medicinal supplements, often consisting of high doses of vitamin A every four to six months; fortification of foods; and nutrition education.

Beriberi and Thiamine Deficiency

Beriberi is a serious disease which was extremely prevalent, particularly in poor rice-eating people in Asia, around the end of the nineteenth century and the beginning of the twentieth. Beriberi, which takes different clinical forms, is caused mainly by thiamine deficiency. Classical cases of beriberi are now reported only sporadically. Because the disease was controlled in the highly endemic areas of Asia some years ago, medical practitioners and public health officials now give less attention to thiamine deficiency and are less familiar than in the past with its manifestations. However, thiamine deficiency leading to a variety of clinical signs, sometimes in conjunction with deficiencies of other vitamins, is not uncommon, but is underreported. Thiamine deficiency is prevalent in chronic alcoholics in industrialized and developing countries, with manifestations different from beriberi.

Causes and Epidemiology

Experimental investigations in Japan, Indonesia and Malaysia led to medical discoveries that proved that beriberi was a deficiency disease, leading to the discovery of its actual cause. Beriberi can be said to be a disease in part caused by new technologies: it became a scourge as the milling industry expanded through out Asia, providing poor people with highly milled polished rice deprived of its thiamine content, at a financial cost no higher than that of home-pounded rice, but at the cost of many thousands of lives. In Asian countries such as China, Indonesia, Japan, Malaysia, Myanmar, the Philippines and Thailand, beriberi used to be a major cause of morbidity and mortality in those whose diet consisted mainly of rice. In contrast, people in many parts of the Indian subcontinent were relatively protected from beriberi because they consumed mainly parboiled rice, which conserves enough thiamine. There have been authenticated cases of beriberi in wheat eaters in the Canadian province of Newfoundland and elsewhere, and also in those consuming other staple foods, but high prevalence rates have been confined to rice-eating people.It has been suggested that an outbreak of disease in Cuba in 1993 may have been caused in part by thiamine deficiency. The manifestations included neurological signs and optic neuritis including loss of sight.

Prevention

People should be encouraged to consume a varied diet containing adequate quantities of vitamin B. If highly milled white rice is the staple diet, part of the rice should be replaced by a lightly milled cereal such as millet, and the diet should be supplemented with foods rich in thiamine such as nuts, groundnuts, beans, peas and other pulses, whole-grain cereals or cereal brans and yeast-based products.

The sale of thiamine-deficient rice and other cereals should be prevented by:

- encouraging the consumption of lightly milled rice and other cereals;
- legislation or other inducement to ensure that all rice put up for sale is lightly milled, parboiled or enriched;
- legislation to ensure vitamin enrichment of cereals made deficient by milling.

Instruction should be given in the most satisfactory ways of preparing and cooking foods to minimize thiamine loss.

Thiamine should be administered in natural food, yeast products, rice polishings or as tablets to certain vulnerable groups in the community.

Nutrition education should be implemented to stress the cause of the disease and to indicate the foods that should be consumed and the ways of minimizing vitamin loss during food preparation.

It is important to strive for early diagnosis of cases of thiamine deficiency and appropriate measures of treatment and prevention.

Thiamine Deficiency in Alcoholics

Although classical beriberi is uncommon in industrialized countries, thiamine deficiency is by no means a rarity. It is prevalent in the alcoholic population in countries both North and South. Alcoholism is an increasingly prevalent condition, and several clinical features previously believed to be due to chronic alcoholic intoxication are now known to be the result of nutritional deficiencies. The most common of these conditions is probably alcoholic polyneuropathy, which has similarities to neuritic beriberi and is believed to result mainly from thiamine deficiency.

Alcoholics who get much of their energy from alcoholic drinks often consume insufficient food and do not get adequate amounts of thiamine and other micronutrients. They may develop a peripheral neuritis, which can influence both the motor and the sensory systems, often affecting the legs more than the arms. The various manifestations include muscle wasting, abnormal reflexes, pain and paraesthesia. These symptoms often respond to treatment with thiamine or B-complex vitamins taken orally.

Another condition resulting from thiamine deficiency in alcoholics is Wernicke-Korsakoff syndrome. Wernicke's disease is characterized by eye signs such as nystagmus (rapid involuntary oscillation of the eyeball), diplopia (double vision arising from inequal action of the eye muscles), paralysis of the external rectus (one of the muscles of the eyeball) and sometimes ophthalmoplegia (paralysis of the muscles of the eye). It is also characterized by ataxia (loss of coordination of body movements) and mental changes. Korsakoff's psychosis involves a loss of memory of the immediate past and often elaborate confabulation which tends to conceal the amnesia. It is now generally agreed that any distinction between Wernicke's disease and Korsakoff's psychosis in the alcoholic patient may be artificial; Korsakoff's psychosis may be regarded as the psychotic component of Wernicke's disease. This view is supported by the fact that many patients who appear with ocular palsy, ataxia and confusion, and who survive, later show loss of memory and other signs of Korsakoff's psychosis. Similarly, psychiatric patients with Korsakoff's psychosis often show the stigmata of Wernicke's disease even years after the illness. Pathological evidence also indicates the unity of the two conditions.

That Wernicke-Korsakoff syndrome is caused by thiamine deficiency and not by chronic alcohol intoxication is shown by the fact that the condition responds to thiamine alone, even if the patient continues to consume alcohol. Of overriding importance in this syndrome is the rapid occurrence of irreversible brain damage; early recognition and treatment are therefore vital. A patient at all suspected of having the syndrome should immediately receive 5 to 10 mg of thiamine by injection, even before a definitive diagnosis is made.

Prevention

The prevention of Wernicke-Korsakoff syndrome calls for considerable public health ingenuity. Several possible measures have been suggested:

— the "immunization" of alcoholics with large doses of thiamine at regular intervals (the development of a suitable depot carrier to reduce the frequency of these injections would be very helpful);

— the fortification of alcoholic beverages with thiamine;
— a provision by public health authorities that thiamine-impregnated snacks be made available on bar counters.

The cost of any of these measures would almost certainly be less than the present enormous expenditure on institutional care of those who have suffered from Wernicke-Korsakoff syndrome.

An optic or retrobulbar neuritis, also known as nutritional amblyopia, that occurred in prison camps during the Second World War was probably caused at least in part by thiamine deficiency not associated with alcoholism. This occurrence may be similar to the serious outbreak of neuropathy disease in Cuba in 1993.

Pellagra

Pellagra, caused mainly by a deficiency of dietary niacin, is generally associated with a maize diet in the Americas, just as beriberi is associated with a rice diet in East Asia.

A number of factors have at different times been suggested as the cause of pellagra. Each theory seemed, when first expounded, to oppose another. Three of the principal theories appear to have an element of truth. Pellagra was first thought to be caused by a toxin in maize, then by a protein deficiency and finally by a lack of niacin in the diet.

It has now been found that maize contains more niacin than some other cereal foods, but it is believed that the niacin in maize is in a bound form. In Mexico, Guatemala and elsewhere where maize has traditionally been treated with alkalis such as lime water to make tortillas and other foods, consumers have been protected from pellagra. It is possible that lime treatment followed by cooking makes the niacin more available, or perhaps it improves amino acid balance. The human body can convert the amino acid tryptophan into niacin; thus a high-protein diet, if the protein contains good quantities of tryptophan, will prevent pellagra. Nonetheless, niacin is still the most important factor in pellagra, and any programme to prevent the disease should aim at providing adequate niacin in the diet. Similarly, all cases of pellagra should receive niacin therapeutically.

Pellagra used to be a very prevalent disease in the southern United States, particularly among poor sharecroppers in the early part of the twentieth century. The disease, unknown in Europe in earlier times, became prevalent in the eighteenth and nineteenth centuries as maize for the first time began to be widely eaten in Italy, Portugal, Spain and parts of eastern Europe. In the twentieth century pellagra has been common in Egypt and parts of southern and eastern Africa, and sporadic cases have been reported in India. In each of these areas the disease was associated with maize becoming the staple diet of poor people who could afford very little else to supplement the diet.

The highest prevalence in recent times has probably been in South Africa, where conditions for some agricultural and industrial workers until 1994 were not unlike those in the southern United States between 1900 and 1920. A report from South Africa suggested that 50 percent of patients seen at a clinic in the Transvaal had some evidence of pellagra, and that the majority of adults admitted to the mental hospital in Pretoria had the disease.

Pellagra regrettably has also been widely reported in refugee camps and in famine situations where maize has been the relief food and relief agencies have given too little attention to providing a balanced diet or adequate micronutrient intakes. An outbreak of pellagra occurred during a drought in central Tanzania in the 1960s when the affected people were consuming mainly donated maize from the United States. The pellagra was quickly controlled using niacin supplements.

Treatment

The following treatment is recommended for pellagra.

- Admission to hospital and rest in bed are desirable for serious cases. Milder cases may be treated as out-patients.
- The patient should be given 50 mg of niacin (nicotinic acid, nicotinamide) three times a day by mouth.
- The diet should contain at least 10 μg per day of good protein (if possible, meat, fish, milk or eggs; if not, groundnuts, beans or other legumes) and should be high in energy (3 000 to 3 500 kcal per day).
- Because the patient may also have a deficiency of other B vitamin components, a vitamin B complex preparation or a yeast product should be prescribed.
- Sedation for a few days is recommended. Those with mental disturbances benefit greatly from any of a number of tranquillizers, for example, valium. The sedative should be given orally, but if the patient is uncooperative more potent tranquillizers may be needed by injection.

Pellagra is often a very gratifying disease to treat. Violent, almost uncontrollable mental patients can become normal, rational, peaceful human beings within a few days of taking a few tablets of nicotinamide. In persons with severe skin lesions, a sore mouth and severe diarrhoea with frequent watery stools, dramatic improvements occur within 48 hours. The skin redness and pain on exposure to sunlight improves; pain in the mouth abates and eating becomes a pleasure for the patient; and most gratifying for the patient, the intractable diarrhoea disappears.

Prevention

The following steps can help in the prevention of pellagra.

— Diversity in the diet is important. Reliance on maize as the sole staple foodstuff should be discouraged, and the consumption of other cereals in place of part of the maize should be encouraged. This is less necessary in those parts of the Americas where maize is treated with lime.

— Production and consumption of foods known to prevent pellagra, i.e. those rich in niacin, such as groundnuts, and those rich in tryptophan, such as eggs, milk, lean meat and fish, should be increased.

— Legislation or other inducement should be put in place to ensure the enrichment of milled maize meal with niacin.

— Niacin tablets should be administered as a prophylaxis in prisons and institutions in areas where pellagra is endemic, and to refugees and in famine relief.

— Nutrition education should be provided to teach people what foods can prevent the disease.

An important lesson to be learned from past experience in the southern United States and current experience in South Africa is that pellagra will be controlled if the conditions for poor agricultural and industrial workers are improved. In the United States the end of slavery, the reduction of sharecropping on southern farms and improvements in wages, working conditions and food supplies had more impact in reducing pellagra than did fortification or medicinal nicotinamide supplements. Recent political changes in South Africa are likely to change and improve the working conditions and diets of poor Bantu in that country and to reduce the prevalence of pellagra there.

Rickets and Osteomalacia

The main feature of both rickets and osteomalacia is a lack of calcium in the bones; rickets occurs in children whose bones are still growing, and osteomalacia in adults who have formed bones. The conditions are, however, caused mainly by a deficiency of vitamin D and not by a dietary lack of calcium. Vitamin D is obtained both from animal foodstuffs in the diet and from exposure of the skin to sunlight. Vitamin D functions like a hormone in regulating calcium metabolism.

Because the body can obtain adequate amounts of vitamin D from even moderate exposure to sunlight, rickets and osteomalacia are uncommon in most African, Asian and Latin American countries, where sunlight is abundant. Where the diseases do occur, they are usually caused in part by a particular cultural practice or local circumstance. For example, in some Muslim societies women practising purdah wear clothes that cover most of the skin, and they and their babies may rarely leave the household. Rickets is reported in some large, densely populated cities (e.g. Calcutta, India; Johannesburg, South Africa; Addis Ababa, Ethiopia), presumably mainly in children who do not get out in the sunlight. Rickets and osteomalacia are now being diagnosed in immigrant families of Asian origin in the United Kingdom. However, nowhere in the tropics or subtropics is rickets a highly prevalent disease, as it was in Europe in the nineteenth century.

Severe rickets usually occurs in children under four years of age who consume only small quantities of foods of animal origin and who for any reason do not have much exposure to sunlight. The bony deformities, however, may be most obvious in older children. Osteomalacia is most common in women who have had several children, who have become depleted of calcium as a result of successive pregnancies and lactation, and who have insufficient vitamin D.

Treatment

Rickets

The basis of treatment is to provide vitamin D and calcium. Vitamin D may be given as cod-liver oil. Three teaspoonfuls three times a day will supply about 3 000 IU, which is adequate. Synthetic calciferol can also be used. Calcium is best given as milk, at least half a litre a day. Cows' milk contains 120 mg calcium per 100 ml.

Tablets containing vitamin D and calcium are available. One of these may be given twice a day to a child under five years of age, and one tablet three times a day to an older child.

While the child is being treated, the mother should be educated regarding the value of sunshine. Rickets, unless severe, is not usually a fatal disease per se, although the child may be more prone to infectious diseases.

Mild bone deformities tend to right themselves with treatment, but in more severe cases some degree of deformity may persist. One of the more serious consequences is obstructed childbirth due to pelvic abnormalities, which may necessitate Caesarean section in hospital.

Osteomalacia

The treatment of osteomalacia is similar to that for rickets. A dose of 50 000 IU vitamin D should be given daily as cod-liver oil or in some other preparation. Calcium should be provided either as milk or, if milk is not available, in some medicinal form such as calcium lactate.

In women with pelvic deformity regular antenatal care is essential, and in some cases Caesarean section before term may be necessary.

Prevention

The prevention of rickets and osteomalacia will depend on the reasons for their occurrence in the particular communities where they are now seen. Usually there is a cultural or environmental cause which may be locally specific and which may need particular attention.

Rickets

Measures should be taken to ensure that all children get adequate amounts of sunlight. In temperate climates such measures include slum clearance; smoke abatement; the provision of parks, playgrounds, open yards and gardens; and regular outings for the young.

Children should have adequate calcium and vitamin D in their diets. Milk and milk products are especially valuable.

Where it is not possible to expose children to adequate sunlight, vitamin D supplements such as cod-liver oil should be given.

Children should attend clinics regularly so that early diagnosis of rickets can be made and curative measures taken.

Nutrition education should be provided regarding the needs for calcium and vitamin D and the methods by which adequate amounts of them can be obtained.

Osteomalacia

The body should be exposed to adequate sunlight. (This need may conflict with religious or social customs, e.g. those requiring women to be heavily covered or veiled, or those forbidding women to go out in public.)

It is important to ensure that a diet containing adequate quantities of calcium and vitamin D is consumed, especially by pregnant and lactating women.

Clinics or home visiting should be established to allow examination of pregnant and lactating women and, where necessary, to issue cod-liver oil or other vitamin D supplements. Advice should be given regarding the consumption of calcium-rich foods. Sometimes medicinal calcium (e.g. calcium lactate) will have to be prescribed.

Nutrition education should be provided, and it should include the topic of child spacing.

Vitamin C Deficiency and Scurvy

Dietary surveys in many countries in Asia, Africa and Latin America indicate that large segments of their populations consume much lower amounts of vitamin C than is considered essential or desirable. Nevertheless scurvy, the classical and serious disease that results from severe deficiency of vitamin C, now appears to be relatively uncommon. No country reports scurvy as a major health problem, but outbreaks are seen in refugee camps, during famines and occasionally in prisons.

Scurvy was first recognized in the fifteenth and sixteenth centuries as a serious disease of sailors on long sea voyages who had no access to fresh foods including fruits and vegetables. Before the era of vitamin research it became practice in the British navy to provide limes and other citrus fruit to prevent scurvy.

Vitamin C or ascorbic acid is an essential nutrient and is necessary for the formation and healthy upkeep of intercellular material; it is like a cement that binds cells and tissues. In scurvy the walls of the very small blood vessels, the capillaries, lack solidity and become fragile, and bleeding or haemorrhage from various sites results. Moderate vitamin C deficiency may result in poor healing of wounds.

Infantile Scurvy

Scurvy sometimes occurs in infants, usually aged two to 12 months, who are bottle-fed with inferior brands of processed milk. During the processing of the milk, the vitamin C is frequently destroyed by heat. Good brands of processed milk are fortified with vita'min C to prevent scurvy.

The first sign of infantile scurvy is usually painful limbs. The infant cries when the limbs are moved or even touched. The child usually lies with the legs bent at the knees and hips, widely separated from each other and externally rotated, in what has been termed the "frog-leg position". Bruising of the body may be seen, although it is difficult to detect in darkly pigmented African skin. Swellings may be felt, especially in the legs. Haemorrhages may occur from any of the sites mentioned above, but bleeding does not take place from the gums unless the child has teeth.

Treatment

Because of the risk of sudden death, it is inadvisable to treat scurvy with only a vitamin C-rich diet. It is advisable rather to give 250 mg ascorbic acid by mouth four times a day as well as to put the patient on a diet with plenty of fresh fruit and vegetables. It is only necessary to inject ascorbic acid if the patient is vomiting.

Increased intake of vitamin C with meals can have a manifest effect on the absorption of iron. In many iron-deficient populations, increasing vitamin C intake will help reduce the incidence and severity of iron deficiency anaemia.

Prevention

Vitamin C deficiency can most easily be prevented in all societies by consumption of adequate amounts of fresh foods, particularly generous intakes of fruits and vegetables, including green leaves. Guavas and various other tropical fruits, for example, are high in vitamin C.

Recommended preventive measures are as follows:

— increased production and consumption of vitamin C-rich foods, such as fruit and vegetables;

- provision of vegetables, fruit and fruit juice to all members of the community, including children, beginning in the sixth month of life;
- provision of vitamin C concentrates if for any reason the previous two measures are not possible;
- improved horticulture, including the provision of village and household gardens, orchards and vegetable allotments in towns and school gardens;
- encouragement of the wide use of edible wild fruits and vegetables known to be rich in vitamin C (e.g. amaranth, baobab fruit);
- action to avoid and discourage the replacement of fresh vegetables, fruit and other foods by canned and pre-served foodstuffs, and encouragement of the greater use of fresh fruit and juices in place of bottled products;
- nutrition education, which should cover the reasons and need for eating fresh foods, and instruction in means of minimizing vitamin C loss in cooking and food preparation.

Zinc Deficiency

Zinc is an essential nutrient and is apparently deficient in the diets of many people in both industrialized and non-industrialized countries. In nutrition journals in the 1990s more is published on zinc and zinc deficiency than on protein-energy malnutrition (PEM). However, zinc deficiency is not claimed as a major public health problem in any country in the world, and no clear disease syndrome is described for zinc deficiency. In Egypt and the Islamic Republic of Iran a condition in males characterized by dwarfism and hypogonadism (poor development of sexual organs) is associated with zinc deficiency. In the United States and elsewhere, low zinc status in children has been associated with retarded growth, poor appetite and impaired sense of taste.

An extremely rare congenital disease known as acrodermatitis enteropathica results in the child's inability to absorb zinc properly. The condition used to be fatal but is now known to respond to zinc therapy. It is characterized by a serious dermatitis, poor growth and diarrhoea.

Laboratory animals on zinc-deficient diets (usually more severely deficient in zinc than any normal human diet) have exhibited anorexia, decreased efficiency of feed utilization, poor growth, depressed gonadal function, impaired immunity, poor healing of wounds and dermatitis. When a zinc-deficient diet was fed to pregnant rats and monkeys, poor behavioural development was observed in their offspring. It is likely that any or all of these signs and symptoms would occur in humans on a very deficient diet, but apparently most human diets provide enough zinc to prevent these more serious manifestations.

It is not surprising that zinc deficiency is often associated with PEM. A diet that is deficient in total amounts of energy and protein is also likely to be deficient in zinc and many other micronutrients. Many children with PEM have low levels of zinc in blood and hair, but these low levels do not prove that their PEM is due to zinc deficiency. A better diet, including more food, would prevent both PEM and zinc deficiency.

Research currently being conducted in several countries may show that in certain populations zinc supplementation can improve poor growth, perhaps by improving appetite which leads to improved dietary intake and better growth. It may also be shown that zinc can improve the functioning of the immune system and in this way reduce morbidity due to infections, thus reducing PEM.

Zinc is present in most foods of vegetable and animal origin. Good sources include flesh from chicken, fish or mammals (pork, beef, mutton), legumes and whole-grain cereals. The United States recommended dietary allowance (RDA) for zinc for an adult is 15 mg daily. It is unlikely that signs of zinc deficiency would arise if intakes were 5 to 8 mg daily, but absorption of zinc, like that of iron, is quite variable. In cases of kwashiorkor and nutritional marasmus treated in hospital, oral zinc supplements may be recommended. Some paediatricians claim that the zinc supplementation speeds recovery, and it can do no harm.

Dental Caries and Fluorosis

Dental caries is not a deficiency disease. However, it is human beings' most prevalent disease and it is one of the most expensive diseases to treat and to prevent. Dental disease is the only disease that a medical doctor is not trained to treat; its treatment is left to a special category of health professionals.

Fluorosis is a condition that arises from excess intake of a mineral nutrient, not a deficiency.

Dental Caries

Dental caries is the medical term for tooth decay, including cavities in teeth. It begins as a loss or destruction of the outer mineral layers of the tooth. Decay tends to be progressive, with loss of minerals and then loss of tooth protein and formation of tooth cavities. The decay may lead to pain, tooth destruction and sometimes infection of surrounding tissues (abscess). Dental caries is an example of an interaction of nutrition and infection.

Three factors contribute to dental caries:

— host factors, namely a susceptible tooth surface;

— the presence of bacterial flora, usually the pathogenic organism Streptococcus mutans, which is cariogenic;

— the presence of a suitable substrate, that is carbohydrate adherent to or between the teeth which allows the bacteria to survive and flourish.

Carbohydrates are broken down and produce organic acids such as lactic acid which lead to demineralization of the teeth. Formerly sucrose was considered particularly culpable. Recent studies have highlighted the fact that caries prevalence correlates well with sucrose consumption in communities where oral hygiene is poor and where fluoride is absent, but not elsewhere. It is now recognized that any fermentable carbohydrate is equally able to lead to dental caries.

Control of dental caries can in theory involve attempts to control or moderate any of the three factors that contribute to the disease. Adequate intakes of fluoride make the tooth surface less vulnerable to caries; mouthwashes can reduce the bacterial presence; and appropriate eating habits can reduce the contact of teeth with sticky carbohydrate, while tooth brushing can remove carbohydrate adherent to the teeth.

In surveys dental caries is assessed by counting the number of decayed (D), missing (M) and filled (F) teeth in each person examined. The total number of D, M and F teeth gives a DMF index. In a survey conducted in the United Republic of Tanzania in 1964, schoolchildren aged six to 14 years had a DMF index of 0.2; that is, an average of one child in five had one affected tooth. In contrast, in a ten-state survey in the United States in 1968, the DMF index for children aged six to 14 years was 7; that is, the average child had seven decayed, missing or filled teeth.

Twenty-five years ago it would have been true to say that dental caries was much more prevalent in the industrialized countries than in the non-industrialized countries. However, with fluoridation of water supplies and toothpaste and other methods of ensuring adequate amounts of fluoride, and with improved dental hygiene and education, dental caries has declined in Western countries. In contrast, with modernization, changing diets and more frequent consumption of fermentable carbohydrates, dental caries has increased in the developing countries, particularly in urban areas of Africa, Asia and Latin America.

Many nutrients are necessary for the development of teeth and their surrounding structures. Vitamin D, calcium and phosphorus, which are important in bone development, are also essential for the development of teeth. Protein and vitamin A are necessary for the growth of teeth, and as has been described, vitamin C is essential for healthy gums. In terms of preventing or reducing dental caries, however, fluoride is the most important nutrient.

In the 1930s it was observed that persons who had access to drinking-water that contained one to two parts per million (ppm) of fluoride had considerably less tooth decay than those whose water supply had much lower amounts of fluoride. It was subsequently found that in areas where the water had very little fluoride, it was possible to reduce the incidence of dental caries by 60 to 70 percent by adjusting the fluoride level of the water to about 1 ppm.

It is now generally agreed that the appropriate amount of fluoride needed in urban water supplies is about 1 ppm, but each city should decide on the level appropriate to its population.

There is no doubt that the fluoridation of water supplies is a public health measure of very great importance. Every physician, dentist and health worker has a responsibility to urge and support fluoridation of the water supply where needed. Fluoridation has been found to be absolutely safe at 1 ppm for people of all ages and in every state of health. Fluoridation is not a form of medication, only an adjustment of the level of a nutrient, like the fortification of bread with vitamins. It is not an infringement of individual rights.

There are substitutes for fluoridation such as pills, drops and fluoridated toothpaste, but none combine the efficiency, practicality, effectiveness and economy of fluoridation for the general public. It should be appreciated that the increased rates of dental caries where water is not fluoridated have the most serious implications for the poor, who cannot afford, or do not have access to, good dental care.

Another means to reduce dental caries is nutrition education to teach parents and children about cariogenic diets and associated risks; education can encourage better dental hygiene, including brushing of teeth and removal of food from between teeth with toothpicks, dental floss or, as is common in much of Africa, a traditional cleaning stick.

In older people the main cause of tooth loss is periodontal or gum disease. This condition usually begins with the formation of plaque (sometimes called tartar or calculus) by bacteria that survive on carbohydrate adhering to the teeth. Plaque between the teeth and close to the gums may lead to secondary infection, to receding and bleeding gums and eventually to loss of supporting bone and loss of teeth. Cleaning teeth, scraping away plaque and chewing fibrous foods help reduce periodontal disease. In a study of poor women in the United States, some 40 percent between 40 and 50 years of age had lost all of their teeth. In contrast, few Africans living in rural areas have had periodontal disease rampant enough to cause such a loss of teeth. Traditional diets are often relatively protective against both dental caries and plaque formation. Western diets are a risk factor.

Fluorosis

In some parts of the world, including certain areas of India, Kenya and Tanzania, natural water supplies contain much higher than desirable levels of fluoride. Intake of water that contains more than about 4 ppm will result in widespread dental fluorosis in the population. In this condition the teeth become mottled and discoloured. At first the teeth have chalky white patches, but soon brownish discoloured areas develop. Fluorosis is not a serious condition, but local people may not like it.

Of more seriousness is skeletal fluorosis, which may result from prolonged intake of water containing high fluoride levels of 4 to 15 ppm. A survey in northern Tanzania revealed a high incidence of fluorotic bone abnormalities in older subjects who normally drank water containing high levels of fluoride. X-ray examinations showed that the bones were very dense or sclerotic and that abnormal calcification was common in ligaments between the vertebra, where tendons link muscles to bones, and in interosseous areas, for example in the forearm. Skeletal fluorosis may cause back pain and rigidity and neurological abnormalities.

Dental Care

In most poor developing countries there are too few dentists to meet the needs of the population. Usually the ratio of dentists per 100 000 people is much higher in the large cities and extremely low in rural areas. Many countries have acknowledged that most of the dental care needed, including diagnosis and treatment such as fillings, extractions and plaque removal, does not need to be provided by a dentist. New Zealand pioneered the use of dental auxiliaries, and now many developing countries train dental assistants or other dental health workers. These workers have much shorter training than dentists and cost much less to employ, but they are fully able to take adequate care of most dental conditions. The few dentists are treated in the same way as other medical specialists; particularly complicated or difficult cases are referred to them. In many countries strong dental associations have opposed the use of dental auxiliaries and have prevented them from doing much of the dental work. This bar is a disservice, especially when even in rich countries like the United States the poor often cannot get adequate dental care.

Other Nutritional Disorders

Nutritional Neuropathies

The nervous system is the communications system within the body, and it is a highly complicated mechanism. If it does not function properly there can be important consequences. The nervous system needs oxygen and nutrients and obtains its energy from carbohydrates. A range of complex enzymes controls its functioning. These enzymes are proteins, and their activities require the participation of a number of vitamins. It is not surprising therefore that dietary deficiencies can cause symptoms and signs indicating impairment or damage to the nervous system.

The B group of vitamins has a special place in relation to the nervous system. These vitamins are commonly found in the outer layers of cereal grains. Milling tends to reduce the quantity of B vitamins in cereal flours. Deficiencies of B vitamins are therefore common, and cases of various neuropathies are likely to increase. For example, an outbreak of a neuropathy in a Tanzanian institution was caused by a change in diet from lightly milled to highly milled maize meal as the main source of energy.

Neuropathies may lead to weakness and pins and needles in the feet, severe burning pains, ataxia, nerve deafness, disturbances of vision, absent or exaggerated reflexes and other symptoms. There is much overlap in the causation of many of these conditions, and classification is difficult.

Vitamin B_6 deficiency secondary to treatment of tuberculosis with isoniazid leads to a polyneuritis. The cause of a recent outbreak of optic neuritis and an epidemic neuropathy in Cuba has not been definitively determined. The outbreak has subsided but was almost certainly the result of a nutrient deficiency, most probably a dietary deficiency of thiamine. Konzo, an epidemic neurological disease, occurs as a result of excessive cyanide intake in those eating toxic cassava.

Riboflavin Deficiency

A dietary deficiency of riboflavin resulting in clinical signs is very prevalent worldwide, in both industrialized and non-industrialized countries. In the United States a ten-state nutrition survey showed poor riboflavin status in over 12 percent of all subjects and in 27 percent of black people examined. In most studies in poor countries riboflavin deficiency is found to be much more prevalent, often affecting 40 percent of the people. As described below, the main clinical features are lesions of the mouth. A deficiency does not cause either life-threatening disease or serious morbidity.

The most frequently seen abnormalities in riboflavin deficiency are angular stomatitis and cheilosis of the lips. Angular stomatitis consists of fissures or cracks in the skin radiating from the angles of the mouth. Sometimes the lesions extend to the mucous membrane inside the mouth. The cracks have a raw appearance but may become yellowish as a result of secondary infection. In cheilosis there are painful cracks on the upper and lower lips. The lips may be swollen and denuded at the line of closure. The lesions may be red and sore or dry and healing.

Glossitis (inflammation of the tongue) occasionally develops involving a patchy denudation, papillary atrophy and so-called magenta tongue. These conditions are not caused exclusively by riboflavin deficiency.

Scrotal dermatitis in males and vulval dermatitis in females have been particularly well described in experimentally induced riboflavin deficiency. The affected skin is usually intensely itchy and tends to desquamate.

Abnormalities in the eyes including redness and vascularization (visible blood vessels), photophobia and lacrimation have been associated with riboflavin deficiency.

A skin condition named dyssebacia may occur near the nose.

Affected persons often have several signs of deficiency at the same time.

In surveys laboratory assessment of riboflavin status has usually been based (as with other water-soluble vitamins) on urinary excretion of the vitamin. A level below 30 μg of riboflavin per gram of creatinine is considered low. Riboflavin status in the individual is better determined by measuring the increased activation of red blood cell (erythrocyte) glutathione reductase. Few hospital laboratories in developing countries are able to do these tests.

Treatment consists of large oral doses of riboflavin for a few days, followed by lower doses which may need to be taken for a long time unless a diet rich in riboflavin is consumed. A dose of 10 mg riboflavin twice a day for one week, followed by 4 mg daily for several weeks, is recommended. Dietary intakes of around 1 to 1.5 mg daily will be protective. Milk is a particularly rich source of riboflavin.

Pyridoxine or Vitamin B_6 Deficiency

A primary dietary deficiency of vitamin B_6 resulting in symptoms of disease is very rare, because even poor diets contain adequate quantities of this vitamin. Pyridoxine deficiency occurs in developing countries mainly secondary to treatment of tuberculosis with the medicine isoniazid. This drug, which is highly effective and can be taken by mouth, was introduced as a treatment for tuberculosis in the early 1950s and became widely used, in part replacing injection of streptomycin which was until then the most common treatment. Despite the development of other medicines, isoniazid is still widely used. Tuberculosis, largely controlled in industrialized countries in the 1970s, is now in resurgence, with drug-resistant cases and cases related to acquired immunodeficiency syndrome (AIDS) worrying public health officials. In many African and Asian countries tuberculosis has remained prevalent and is an important cause of morbidity and mortality.

Isoniazid taken in large doses over long periods is very likely to precipitate vitamin B_6 deficiency. It is said to increase vitamin B_6 needs.

The deficiency is usually manifested by neurological abnormalities, including a peripheral neuritis which may involve severe pain in the extremities, including the legs. Experience in East Africa showed that because of the pain rural patients were often unable to walk to health centres for examination or to obtain their medicine.

It is strongly recommended that tuberculosis patients being treated with isoniazid be given 10 to 20 mg pyridoxine by mouth daily. Unfortunately, pyridoxine is much more expensive than isoniazid, so providing the vitamin greatly increases the cost of treatment.

It has been suggested that in certain parts of the world, particularly in Thailand, low intakes of vitamin B_6 may be responsible for bladder stones. It is known that vitamin B_6 increases oxalate excretion in urine and that vitamin B_6 deficiency leads to an increased risk of oxalate stone formation in the kidney or bladder.

Hormonal contraceptive pills have been associated with both folate and vitamin B_6 deficiencies. However, the newer birth control pills have not been shown to result in vitamin B_6 deficiency. Oral vitamin B_6 tablets have been claimed to reduce the nausea of some women in the first months of pregnancy.

An extremely rare congenital disease called pyridoxine-responsive genetic disease leads to hyperirritability, convulsions and anaemia in the first few days of life. Unless treated very early with vitamin B_6 the child develops serious permanent mental retardation.

Chronic diseases with Nutritional Implications

In the relatively wealthy industrialized countries most nutrition research, teaching and action relates to certain chronic diseases in which diet has a role. These include obesity, arteriosclerosis and coronary heart disease, hypertension or high blood pressure (which may lead to stroke), certain cancers, osteoporosis, dental caries and tooth loss, some liver and kidney diseases, diabetes mellitus, alcoholism and other diseases. Most of these diseases have known dietary or nutritional factors in their aetiology or in their treatment, or in both. It is now evident that the incidence of many of these chronic diseases or conditions is increasing in developing countries, especially in more affluent sections of their populations. The implications of the transition or of the coexistence of different nutritional conditions in two parts of the population present significant public health problems for these nations. It is important that the countries consider appropriate agricultural, public health and other policies that could mitigate or even avert the bad effects of these changes.

It is striking that in the United Kingdom between about 1942 and 1947, when very strict rationing was imposed as a result of the Second World War, the British people were probably better nourished than ever before or after. Severe restrictions were put on each family, particularly regarding the amount of meat, butter, eggs, edible fat and other foods of animal origin in their diets. Fruits and vegetables were not rationed. Rationing applied to both the rich and the poor, and it is believed to have been rather fairly implemented. The rich certainly reduced their intake of foods of animal origin, and the poor received their fair share. Both groups of the population benefited nutritionally. Even mortality rates from diabetes were markedly lowered.

Rationing of foods is not suggested as a strategy in normal times. However, the British experience suggests that more equitable consumption of certain foods may be nutritionally beneficial to both segments of the population, reducing both undernutrition and overnutrition.

It has been recognized that excessive intake of energy, certain fats, cholesterol, alcohol and sodium (mainly salt) and decreased consumption of fruits, vegetables and fibres coupled with sedentary lifestyles contribute importantly to increased incidence of chronic diseases in the affluent sectors of most communities around the world. These diseases are often described as nutritional diseases of affluence, which is an easy but misleading description. Factors other than income influence the changing incidence of these diseases, and in many more affluent countries it is the poor who suffer most from these diseases.

This section provides a brief discussion of the causes, manifestations and prevention of some of the more important nutrition-associated chronic diseases:

- arteriosclerotic heart disease,
- hypertension or high blood pressure,
- diabetes mellitus,
- cancer,
- osteoporosis,
- other conditions.

In some of these diseases the cause is clearly dietary; in others diet may be important in contributing to the cause or in treatment; and in some the relationship to diet is suspected but not proven.

Arteriosclerotic Heart Disease

Coronary heart disease caused by arteriosclerosis is one of the leading causes of deaths in most industrialized countries in North America, Europe and elsewhere. Over half a million people die of arteriosclerotic heart disease in the United States each year. Working in three different rural hospitals in the United Republic of Tanzania in the 1960s, the author did not see a single case of coronary thrombosis in an African patient. Arteriosclerotic disease is associated with many risk factors which appear to be common in middle-aged and older men and post-menopausal women living in industrialized countries of the North; they are apparently much less common in traditional rural societies in countries of the South. The situation is changing, however, and heart disease and stroke are becoming important causes of mortality in many Asian and Latin American countries.

Causes

The actual cause of arteriosclerosis and the coronary thrombosis that may result is not exactly known. Various factors lead to deposits of lipid material in the arteries. The deposits may at first be lipid streaks, but they may then be followed by atheromatous plaques and often a narrowing of the coronary arteries.

Even if the exact cause of arteriosclerosis is not known, risk factors that increase the likelihood of serious arteriosclerosis are recognized:

— Hypertension or high blood pressure adds to the risk of serious arteriosclerosis and coronary thrombosis (as well as stroke).

— Raised serum lipids (high levels of serum cholesterol and low levels of high-density lipoproteins) are strongly associated with arteriosclerosis.

— Cigarette smoking is an important risk factor; several studies have shown a marked increase in coronary thrombosis and other manifestations of arteriosclerosis in those who smoke cigarettes compared with those who do not.

— Diabetes mellitus is well recognized as a risk factor in arteriosclerosis.

— Hormonal levels have a role. There is little doubt that up to about age 45 females are at much lower risk of arteriosclerosis and coronary thrombosis than males, but after menopause the differences narrow or disappear. Although it has not been proved, oestrogen appears to protect from coronary heart disease and testosterone may increase the risk.

— Lack of exercise is a factor. Sedentary people are more likely than active individuals to get arteriosclerosis.

— A genetic predisposition for the disease is a possibility; certain persons seem at greater risk, probably because of genetic influences or a family predisposition.

Of all the causative or risk factors that can be realistically manipulated to reduce arteriosclerosis, nutritional factors and smoking stand out as deserving of action. In experiments in animals dietary manipulation has been the easiest way to stimulate arteriosclerosis.

Mean levels of blood lipids and serum cholesterol in humans differ greatly between countries with high rates of heart disease mortality and those with low rates. Lipoproteins are classified as very low density lipoprotein (VLDL), low density lipoprotein (LDL) and high density lipoprotein (HDL). HDL is often termed "good cholesterol" and LDL "bad cholesterol". Increased rates of heart disease are associated with high levels of LDL, so high LDL levels indicate increased risk. In contrast, HDL may be protective against arteriosclerosis and low HDL levels increase risk. An LDL/HDL ratio above 3.5 indicates a high risk.

Total cholesterol concentration below 5.2 mmol/litre is interpreted as low risk of coronary heart disease, between 5.2 and 6.2 mmol/litre as moderate risk and above 6.2 mmol/litre as high risk. However, risk is also influenced by other risk factors, such as smoking.

Prevention

In general, people could take the following dietary and related steps to reduce the likelihood of getting coronary thrombosis.

— Ensure that energy obtained from fat constitutes less than 30 percent of total energy intake (35 percent if the person is active) and that less than 10 percent of energy comes from saturated fat; increase the proportion of fat coming from polyunsaturated fat.

— Consume less than 300 mg of dietary cholesterol per day.

— Consume food that provides energy in appropriate amounts to ensure desirable body weight while undertaking a healthy level of physical activity.
— Consume less than 10 μg of salt per day. (This step is likely to help reduce hypertension - a condition associated with arteriosclerosis - among salt-sensitive individuals.)
— Avoid cigarette smoking.
— Maintain optimum body weight, and if obese lose weight.
— Treat and control diabetes if present.
— Maintain blood pressure in the normal range.

Some scientists also recommend a high intake of antioxidant vitamins, particularly vitamin C and beta-carotene but also vitamin E, to reduce the risk of arteriosclerosis and some cancers.

In view of the above, the practical dietary guidance would be to maintain energy balance and to ensure adequate intake of fruits, vegetables, legumes and grains.

In recent years several industrialized countries have reported that deaths from coronary heart disease have decreased parallel with dietary changes, particularly reduced intakes of certain fats and oils and increased consumption of fruit, vegetables and fibre. The changes have partly come about because the public has been educated and informed about diets and other lifestyle factors that may contribute to heart disease and because as a result the food industry has changed certain practices in response to consumer demand. Thirty years ago low-fat milk was hardly used in the United States; now milk that is skimmed or 1 or 2 percent fat is widely available, and the majority of Americans use non-fat or low-fat rather than whole-fat milk.

Obesity

Obesity is often considered a condition of affluence. Certainly in affluent nations such as the United States obesity is highly prevalent, and in most poor African countries it is much less common. However, obesity or overweight is common in both adults and children even among the poor in some non-industrialized countries, particularly the middle-income nations. In several Caribbean countries over 20 percent of women are classified as obese.

Obesity, especially severe obesity, is associated with high risks of coronary heart disease, diabetes, hypertension, eclampsia during pregnancy, orthopaedic problems and other diseases. Obesity has been found to be associated with excess mortality.

Causes

When over a prolonged period more energy is ingested in food than is expended by physical exercise, work and basal metabolism, weight will be gained and obesity will result. Metabolic studies show that diets high in fat are more likely to induce body fat accumulation than diets high in carbohydrate. There is no evidence that simple sugars differ from complex sugars in this regard. High intake of dietary fat is positively associated with indexes of obesity.

Obesity is only rarely due to endocrine (glandular) dysfunction. A very small amount of food energy consumption above energy expenditure is enough to lead to obesity a few years later. Consuming 100 kcal more than needed per day (one slice of bread and butter, 100 µg of maize porridge, 220 µg of beer, 26 µg or a little more than two tablespoonfuls of sugar) would lead to a gain of 3 kg per year, or 15 kg in five years.

While obesity is due to an imbalance between energy intake and energy expenditure, other underlying causes - a metabolic condition, endocrine disorders or genetic factors - may certainly contribute.

Among affluent people obesity may be in part due to a tendency to take less exercise and do less energetic physical work than less affluent people. Poor rural people who engage in agriculture and walk long distances burn up much energy because of their high exercise level. When rural people move to urban areas and become more affluent they may have less need to exert energy or to do heavy physical work, and they may have access to more food and more energy-dense food, which may contribute to obesity. Obesity may become a vicious cycle, because an obese person may have more problems than others in walking long distances or in doing heavy physical work.

Control of obesity

Because the treatment of obesity is difficult and often fails, prevention of overweight is preferable to treating the problem after it has developed. Nutrition education, starting in schools, can provide persons with the information, and perhaps the motivation, always to balance energy intake with energy expenditure. Maintaining a high level of activity is helpful. In developing countries, especially in rural areas, there is no need to institute programmes of jogging or aerobic exercises. Rather it is important to value physical work and to encourage all people of all ages to do an appropriate amount of physical work, be it labouring in the fields, chopping wood in the home or public service activity; to walk where feasible rather than use alternative transport for short-distance journeys; and, if desired and feasible, to indulge in sport.

Some health professionals would recommend that treatment is warranted only for Grade II and Grade III obesity. People with BMI between 25 and 29.9, if it remains in that range, do not have much added risk of disease or reduced life expectancy. Nevertheless, all obese people have passed through Grade I to reach Grade II and Grade III. Thus for Grade I subjects aggressive treatment is not called for but prevention is; these people should take steps not to become more obese.

The only logical way to treat obesity is to reduce energy intake and increase its expenditure. Energy intake can be lowered by reducing the size of servings at each meal; energy expenditure can be raised by increasing the amount of exercise taken. However simple this may sound,- long-term maintenance of lowered weight is very difficult for people who have been obese.

Recent studies suggest that energy balance is maintained under free living conditions if a balance between intake and oxidation is also reached for each macronutrient (carbohydrate, protein and fat). In the cases of protein and carbohydrate, oxidation normally matches intake. Fluctuations in energy balance are therefore mainly governed by variations in fat balance. In the context of weight reduction this means that to induce a negative fat balance, daily fat oxidation must exceed daily fat intake. Regular prolonged exercise and reduced intake of fat would thus result in substantial weight and fat loss. In the end a new fat balance is achieved by the human body at a reduced body fat mass. Therefore, the best way to reduce the energy intake of the diet for weight reduction is to cut down fat intake and increase the intake of vegetables and fruits.

There is no prophylactic treatment that will of itself induce weight loss. The use of amphetamines, thyroid extracts and other drugs in the treatment of obesity is in general to be condemned and at best should be carefully supervised by an experienced doctor. Similarly, most of the highly advertised rapid-reducing diets, of which some are even promoted by physicians, have been found to be ineffective and sometimes dangerous.

Diabetes Mellitus

Diabetes mellitus is a chronic metabolic disorder in which blood glucose levels are raised because of a deficiency or diminished effectiveness of insulin. The disease is not curable, and it may lead to a variety of complications, some of them serious. Treatment can reduce the complications. Diabetes is occasionally secondary to other diseases, especially those that affect the pancreas, the organ that produces insulin.

For a long time it has been known that diabetes occurs in families, and that therefore there are genetic factors involved. However, families also usually share an environment, eat similar foods and have a common pattern of activity Dietary factors and activity pattern have a role, and in Type 2 diabetes obesity is a frequent precursor. Obese diabetics who lose weight improve their condition. There is no evidence that large intakes of sugar increase the likelihood of diabetes or that diets high in fibre and complex carbohydrates reduce the likelihood of diabetes except insofar as they displace fat in the diet and lower the risk of obesity. Type 1 diabetes in some cases appears to be associated with early viral infections.

The report of the International Conference on Nutrition (FAO/WHO, 1992a) suggests that an "apparent epidemic of diabetes is occurring in adults 30 to 62 years of age throughout the world" and that the trend is "strongly related to lifestyle and socio-economic change". The trend concerns mainly non-insulindependent or Type 2 diabetes. For this age group levels of diabetes are moderate, between 3 and 6 percent, in Europe and North America and in some developing countries. A high prevalence (10 to 20 percent) is seen in some urban Indian and Chinese societies and in immigrants (sometimes second or third generation) from the Indian subcontinent who have settled in the Caribbean, Fiji Mauritius, Singapore and South Africa. Diabetes is uncommon in many communities in the developing world where traditional diets and activity patterns are maintained.

It is not absolutely clear why particular migrant groups or others changing their lifestyles from traditional to sedentary seem to be at special risk for diabetes. It seems highly likely, however, that dietary changes, sometimes including excess alcohol intake, are a major factor. The dietary changes are also accompanied by an altered way of life, from rural to urban; from hard physical work to a sedentary life; and possibly from rural poverty to somewhat greater affluence.

Treatment and control

The aim of treatment is to maintain health and to avoid complications. This is achieved by trying to get blood glucose levels as close to normal as possible for as much of the time as possible, and by so doing to reduce the amount of glucose spilling into the urine. Control is greatly assisted by reduction of weight in obese diabetics and by maintenance of a healthy body weight in all diabetics.

There are three cardinal principles in the treatment and control of diabetes: discipline, diet and drugs. Diabetics must organize a regular and disciplined lifestyle, with timely eating, work, recreation, exercise and sleep. They must regulate their food intake to meet their diabetic requirements and use drugs as a recourse only when this regimen fails to control the condition. Control requires good cooperation between the sufferer and the health worker and an understanding that there is no cure but that often good health can be maintained into old age. Most Type 2 diabetes can be controlled by discipline and diet. Many young Type 1 diabetics and a few more serious Type 2 diabetics may need insulin or other drug therapy under close medical supervision. Elderly diabetics are often overweight, and their diets need to be adjusted to help them achieve desirable weight. This is feasible but not easy.

There is still debate and disagreement about the best dietary treatment for diabetes. Readers should consult comprehensive textbooks of nutrition or internal medicine for detailed advice. Many physicians now recommend a diet in which 55 to 65 percent of the energy comes from carbohydrate, 10 to 20 percent from protein and 20 to 30 percent from fat. The diet should be mixed and varied, containing cereals, legumes or root crops, fruits and vegetables. Foods high in fibre are desirable.

What is important is that feeding be regular. The diabetic should eat modest amounts frequently and avoid either bingeing or going for long periods without eating. Dietitians often find it useful to provide exchange lists which inform the diabetic about groups of foods or dishes that contain similar amounts of carbohydrate, protein, fat and energy.

Osteoporosis

Osteoporosis is a chronic disease that is now very common in older people, particularly women, in industrialized countries. The disease is characterized by excessive demineralization of the bones of the skeleton. In general, reduction in the calcium content of the bones has been considered a normal ageing process. Particularly in post-menopausal women in industrialized countries, however, the loss of bone density is accelerated.

Osteoporosis greatly increases the risk of fractured bones, even from falls or minor trauma. Fractures of the neck of the femur (near the hip joint) are almost epidemic in female senior citizens in North America and Europe, and these people also frequently have fractured vertebrae. They may become shorter, have bent backs and suffer excruciating pain.

The cause of osteoporosis is not known. Almost certainly in females it is due in part to lower levels of female hormones (such as oestrogen) after the menopause and to taking little exercise. Some believe that low calcium intakes have an important role, and many millions of people take medicinal calcium with the belief that it will reduce their chances of getting osteoporosis. However, dietary intakes of calcium are much higher in North America, where osteoporosis is prevalent, than in many countries in Asia and Africa where osteoporosis is uncommon. High protein intakes increase the need for calcium, so Western people consuming high-protein diets do have increased calcium requirements.

There is some evidence that increasing intakes of fluoride helps to maintain bone density, and fluoride in the past was tried in osteoporosis treatment, but it is not now widely recommended. Many women in industrialized countries now take oestrogen after the menopause, and this probably reduces the demineralization leading to osteoporosis. Regular relatively strenuous exercise also reduces the loss of bone density. Rural women in Africa, Asia and Latin America, who as long as they are fit work in the fields, carry wood and water, walk long distances to the market and in general are highly active, seem to do what is needed to lessen their likelihood of developing osteoporosis. Immobilized humans, be they fracture patients confined to bed or astronauts in space, definitely lose calcium from their bones.

In North America and Europe increasing intake of calcium may reduce the likelihood of developing osteoporosis. In the United States and the United

Kingdom milk provides 30 to 50 percent of the dietary calcium consumed. Whole milk, if consumed in the quantities often recommended to prevent osteoporosis, will also markedly increase, possibly to unhealthy levels, intakes of total fat, saturated fat and energy. Calcium supplements are often recommended. Recent experimental evidence in humans suggests that treatment with parathyroid hormones may be effective in some cases of osteoporosis.

Nutritional Problems of Poverty and Affluence

Many of the main deficiency diseases prevalent in developing countries are associated with food insecurity, poverty, infectious diseases, inadequate care and related factors. It has been clearly shown that so-called economic development, especially development that goes hand in hand with poverty alleviation, leads quite rapidly to major reductions in malnutrition and infections. Examples of countries where this has been the case include Costa Rica and Cuba in Latin America; Malaysia and Thailand in Asia; and Mauritius in Africa. The major reductions in malnutrition, in prevalence of communicable diseases and in infant and child mortality rates are probably usually a result of improved education and reduction in illiteracy, greater household food security, better hygiene and water supplies and wider access to reasonably good health services.

In most countries as the rates of protein-energy malnutrition and of infections such as gastro-enteritis and intestinal parasites go down, there is often an increase in the incidence of arteriosclerotic heart disease, obesity, certain cancers, diabetes and stroke. The transition and the changing health profile are often first evident in the most affluent and in urban rather than rural populations.

Reliable morbidity data are frequently not available, but in many countries mortality data are published. These data clearly show that in many of the better-off developing countries deaths from infections and malnutrition have markedly declined and infant mortality rates have greatly improved. However, the mortality rates from what are termed "diet-related non-communicable diseases" have increased in these nations. These diseases include malignant neoplasms, diabetes, obesity, circulatory system diseases (excluding rheumatic fever), chronic liver disease and cirrhosis, cholelithiasis and cholecystitis. World Health Organization (WHO) statistics for 42 countries which had good mortality data for 1991 to 1992 (WHO, 1993d) show that in some industrialized countries such as Australia, Japan, the United Kingdom and the United States the mortality rates from these causes decreased from 1960 to 1990, while in more affluent developing countries such as Ecuador, Mauritius and Thailand the death rates from these causes markedly increased over the same period. In many of these middle-level developing countries the death rates from these diseases in persons 45 to 54 years of age were very similar to those of industrialized countries for the period from 1985 to 1989. It is likely that the significant decreases in the industrialized countries are due to educational efforts and public health messages that influence people to reduce their dietary intake of harmful dietary components and to change behaviours that increased the risk of dying from these disorders. Certainly non-nutritional behaviour changes, for example reduced cigarette smoking, also contribute to these reductions. The dietary change usually believed to be most important is a reduction in the consumption of certain fats.

An increase in the diet-related noncommunicable diseases in rapidly developing countries is likely first to affect more affluent people, often productive well-educated persons in important positions in both the public and private sectors. These diseases then may reduce the productivity of these people, and their treatment may also begin to absorb a larger and larger segment of the health care budget. The challenge for nutritionists and others is to help emerging developing countries avoid the transition from a high prevalence of preventable infections and malnutrition to increasing rates of partly avoidable chronic diseases of affluence.

The developing countries, especially those that are rapidly industrializing and witnessing rapid increases in incomes, are in a position to take action before major increases in these diseases occur. This is a challenge that should be grasped and not ignored. Perhaps measures to reduce cigarette smoking are even more important than those to prevent harmful changes in food intake, but actions to prevent harmful dietary practices deserve priority. China is one country that is at least considering these problems and appropriate actions. Its attention to these problems is particularly important because China is the world's most populous country and has transformed itself in 50 years from a country with much extreme poverty, severe food shortages and many deaths from infections to a nation with a booming economy, food security and a health service that has controlled many preventable infections. The Chinese Government has a good deal more control over its citizens than do many other governments, and it could take steps to reduce the already rising rates of nutrition-related and cigarette-related chronic diseases. In so doing, China could set an example for other countries.

In the mid-199Os concern is now focused on the emerging problem of cardiovascular disease in the countries of Eastern Europe and the former Soviet Union. The increasing incidence of chronic diseases in the developing countries deserves attention.

Dietary Guidelines

Guidance in nutrition can have various purposes. It can be provided for the setting of national priorities in the health sector, or to facilitate the planning of national economies (dietary goals, dietary/nutritional targets); or it can address individuals (recommended nutrient intakes, dietary guidelines). All these forms of guidance have in common the goal of helping populations achieve a state of optimal nutrition which is conducive to good health.

Since human beings everywhere have rather similar nutrient requirements relative to their age, gender and body size, nutritional guidance can be prepared in a global perspective to some extent. Strategies for achieving nutritional goals, however, will vary from one population to another; they will need to take into account the biological and physical environment of the population, as well as economic and relevant socio-cultural factors. These aspects should be reflected in dietary guidelines.

Dietary guidelines are sets of advisory statements providing principles and criteria of good dietary practices to promote overall nutritional well-being for the general public. They are intended for use by individuals.

Dietary guidelines are primarily based on the current scientific knowledge regarding nutritional requirements and also, indirectly but strongly, on the types of diet-related diseases prevalent in the given society. The guidelines take into account the customary dietary pattern and indicate the modifications that should be made to contribute to the reduction of these diseases. They represent the practical way to reach the overall nutritional goals for a population.

Until recently, dietary guidelines have usually been expressed in technical nutritional terms. Now, however, food-based dietary guidelines, which express the principles of good dietary practice in terms of foods, are increasingly common. Where they cannot be expressed entirely in terms of foods they are written in ordinary language. These guidelines avoid as far as possible the technical terms of nutrition science. Food-based dietary guidelines vary among population groups. Hence it is important that each region or country recognize that more than one dietary pattern is consistent with health and develop food-based strategies that are appropriate for the local region.

Food and diet are not the only components of a healthy lifestyle. Therefore, organizations developing dietary guidelines are encouraged to integrate diet-related messages with other policies related to health (e.g. smoking, physical activity, alcohol consumption).

The following key points should be considered in the formulation of dietary guidelines:

- Public health issues should determine the direction and relevance of dietary guidelines.
- Dietary guidelines address a specific socio-cultural context and therefore need to reflect the relevant social, economic, agricultural and environmental factors affecting food availability and eating patterns.
- Dietary guidelines need to reflect food patterns rather than quantitative goals.
- Dietary guidelines need to be positive and should encourage enjoyment of appropriate dietary intakes.
- Various dietary patterns can be consistent with good health.

To address better the issues of optimal nutrient intakes for the development of food-based dietary guidelines, the recent FAO/WHO Consultation on Preparation and Use of Food-Based Dietary guidelines (1995) advocated the concept of nutrient density applied to total diet - i.e. the amount of essential nutrients provided per 1000 kcal of energy provided by the diet - as an alternative to the traditional focus on recommended dietary allowances (RDAs) for specific nutrients.

Get the Best from Your Food - An FAO Initiative

FAO has recently produced a set of nutrition education materials which are based on the above considerations and can facilitate the development of practical dietary guidelines. The package, entitled Get the best from your food, is based on the recognition that food has value and significance well beyond the supplying of nutrients. Eating is among the most natural and pleasurable activities known, and within society food, and especially the sharing and securing of food, has considerable social significance. The multiple roles of food and eating behaviours need to be recognized and appreciated in the development of dietary guidelines.

The FAO initiative is based on four principles:

— The human body is a very adaptable organism, and a wide variety of dietary patterns and food intakes can lead to good health and nutritional well-being.
— From a nutritional perspective, a given food can neither be required nor proscribed. There are no good or bad foods, per se, only good and bad diets.
— Diets, in themselves, can only be judged good or bad in relation to a number of other variables, ranging from an individual's physiological status to physical activity levels, lifestyle choices and environmental conditions. Helping consumers understand what these variables are and how they can be modified beneficially is a major objective of dietary guidance.
— Dietary intake, except in extreme situations, is primarily a matter of choice, and dietary guidance can be most effective in helping people make good food choices through positive, non-coercive messages.

Messages of Positive Dietary Guidance

Get the best from your food initiative is based on four messages that can be used to develop not only dietary guidelines, but also educational programmes for public information, schools and other training settings. The concept and messages are positive, simple and direct. They are intended to promote healthful and realistic consumption patterns among all age groups and to encourage sound, practical approaches to food and nutrition.

Enjoy a variety of food. This message embodies two concepts. The first is that food, eating and dietary guidance need to be seen in a positive light. This idea is especially important given the negative messages often associated with dietary guidance, especially in more affluent societies.The second concept is that dietary adequacy must be based on dietary diversity. This message stresses that the consumption of a wide variety of foods is necessary and that all types of food can be enjoyed as part of a wholesome diet. Recognizing the benefits of mixed and varied diets is especially important in light of the still incomplete understanding of nutritional requirements, nutrient and non-nutrient interactions and diet-health relationships.

Eat to meet your needs. This message emphasizes the changing nutritional needs throughout the life cycle and how those needs can best be met from locally available foods. Attention is drawn to energy and nutrient requirements during high-risk periods (pregnancy, lactation, infancy, illness, old age) and in difficult situations such as times of low food availability. This message also permits problems associated with overconsumption and unbalanced dietary intakes to be addressed.

Protect the quality and safety of your food. This concept is often overlooked by those providing dietary guidance, yet it is of great importance in both developed and developing countries. In many developing countries malnutrition is often caused by the poor state of water and food sanitation, and in all countries the consumption of poor-quality, contaminated foods is a major health risk. Vigorous efforts to protect the quality and safety of food supplies need to be made within households, schools and other institutions and at village-level and commercial processing and storage facilities.

Keep active and stay fit.. This message emphasizes that nutritional well-being is not just a matter of eating properly. Human bodies need exercise to function well and stay healthy. Many of the diet-related chronic diseases are closely linked to activity patterns, and efforts to improve nutritional well-being need to consider this fact.

References

Alleyne, G.A.O., Hay, R.W., Picou, D.I., Stanfield, J.P. & Whitehead, R.G. 1977. *Protein-energy malnutrition.* London, UK, Arnold.

Berg, A.1987. *Malnutrition. What can be done? Lessons from the World Bank experience.* Baltimore, Maryland, USA, Johns Hopkins University Press.

Waterlow, J.C. 1992. *Protein energy malnutrition.* London, UK, Edward Arnold.

World Bank.1994. *Enriching lives. Overcoming vitamin and mineral malnutrition in developing countries.* Washington, DC, USA.

5

Nutritional Assessment

Nutritional problems are complex in their aetiology, and there are many different nutritional deficiency diseases Knowing how they occur is one vital part of solving and, better still, preventing nutritional problems The ability to predict their occurrence makes prevention a more realistic prospect

A great variety of data can throw light on the risks of malnutrition in a community or a nation Between 1946 and 1975 large national nutrition surveys were conducted in many countries They often included the collection of a broad range of dietary, clinical, biochemical, anthropometric and socio-economic data The surveys were often designed to detect evidence of a range of vitamin and mineral deficiencies as well as protein-energy malnutrition (PEM) The surveys were expensive to conduct; they required well-equipped laboratories and numerous personnel Many of the earlier surveys in over 20 countries were supported and largely conducted by the United States Interdepartmental Committee for Nutrition for National Defense Subsequently, international agencies such as FAO helped countries conduct large national nutrition surveys. In the United States, major nutrition surveys were conducted in ten states between 1968 and 1971.

All of these surveys provide a wealth of data on nutritional status, usually for a representative sample of the population. Unfortunately, in most cases the data collection did not seem to result in a broad set of actions to deal with the nutritional problems found in the surveys.

By about 1975 it was generally agreed that such detailed surveys were not necessary and that, because PEM in young children was thought to be the most important problem, simplified surveys, using mainly anthropometry and selected dietary and socio-economic indicators, would be more appropriate. Nutritional assessments were increasingly based on measurements of weight and height. There also was a move away from national surveys to more local surveys and in some countries, such as Kenya, to regular data collection to assess trends. Anthropometric surveys were to some extent replaced in the 1980s by rapid appraisal methods which involved the collection of a broader range of data but used new methodologies. At about the same time there was a move to collect qualitative as well as quantitative data and to conduct surveys related to a single micronutrient deficiency, such as iodine deficiency disorders (IDD).

In working to assess the nutritional status of a community, it is important to decide on the objectives of the assessment, how the analyses will be done and what actions are feasible. It is important to draw from experience and to design the most appropriate data collection exercise. For example, in an assessment in a large, newly established refugee camp, it might be advisable to collect more than just anthropometric data; in the past, when nutritional status in refugee camps was judged only on anthropometry, deficiency diseases such as scurvy and pellagra were missed. Social scientists might be consulted to help decide what qualitative data would be most useful and how these might be gathered and analysed.

Large and expensive surveys, in which a wide variety of nutrition-related data are collected, are seldom justified and should never be done unless there is reasonable assurance that the data will be used for an action programme and that adequate resources and funds are available. In many countries expensive surveys have been carried out and little action has followed. It has been suggested that ten times the amount spent on a survey should be available for programmes aimed at overcoming the deficiencies identified by it. It is therefore important that the information collected be kept to the minimum required to assess or monitor the situation, and that surveys be simplified as much as possible. Some information used for the assessment of the nutritional status of a community can also be used for evaluation of programmes and for nutritional surveillance.

Today the main interest in a survey might be to determine nutritional status at the household and local level, rather than at the national level. The following ten types of information can be useful in assessing the nutritional status of a community:

— clinical examination data;
— anthropometric data;
— laboratory tests of nutritional status;
— dietary surveys;
— vital statistics;
— additional health statistics and medical information;
— food availability and market surveys, including agricultural data relevant to food production and food balance sheets;
— economic data related to purchasing power, food prices, food distribution, etc.;
— socio-cultural data, including food consumption patterns and food practices and beliefs;
— food science information such as the nutrient content of foods, the biological value of diets, the effects on nutrients of common food processing practices and the presence of toxic or harmful factors such as aflatoxins and goitrogens.

Only the first five are discussed here since a nutrition survey comprehensive enough to collect all these types of information would very seldom be undertaken.

Clinical Examination

Clinical examinations are often given low priority as a means of assessing the nutritional status of a community. Moreover, most countries in Africa, Asia and Latin America suffer a lack of vital statistics, accurate figures for agricultural production and laboratories where biochemical tests can be performed. Records of local food habits and practices are difficult to obtain. Under these conditions clinical and anthropometric examinations are the most simple, most practical and without doubt the most sound means of ascertaining the nutritional status of any particular group of individuals.

The nutritional status of a community is the sum of the nutritional status of the individuals who form that community. However, in any survey only a representative group of persons needs to be examined. To give a true picture, these people should normally be chosen completely at random, not taken from any particular age group, sex, religion, social class or area within the community. Stratified sampling is valid under certain circumstances. For example, if a survey is being carried out to determine the importance and prevalence of PEM among the young in a given area, it would be sound to restrict examinations to children up to five years of age. If the exact date of birth of the child is unknown, the age should be estimated using local historical, agricultural or social events as time indexes.

The clinical nutrition examination should be carried out by a person with medical training. Although it may be possible to train non-medical personnel to recognize such conditions as angular stomatitis, mottled teeth and even oedema, collection of clinical data by people with inadequate medical knowledge could lead to incomplete survey results. For example, a person looking for the dermatoses of kwashiorkor or the skin changes of pellagra should also be able to recognize scabies and eczema. However, non-medical persons can be entrusted to collect anthropometric data (physical measurements).

In order to avoid overlooking important details, the clinical examination should be systematic. The examiner should look for specific signs, and their presence or absence should be recorded on a standardized form. A modified sample of a form that has been found useful in East Africa is presented on the following page.

Using this form, examinations should start at the head (i.e. hair, eyes, mouth), move down the body and end at the feet. Central nervous system (CNS) signs may in some instances be omitted; they are relatively rare and the tests may be difficult and time consuming to perform.

ANTHROPOMETRIC DATA

Anthropometric data can be collected by medical or non-medical personnel. In the former case, they can be included as part of the clinical nutritional examination. However, it is often simpler and faster if a reliable person other than the medical examiner records the height and weight during a survey.

Weight

The weight of a person is the most important single anthropometric measurement that can be taken. In children its interpretation is dependent on knowing the age of the child with some degree of accuracy. Weight should be measured with the subject nude or wearing the minimum of clothing (shorts only for males, light dress for females). Footwear should be removed.

Spring scales are less reliable than balance scales. In many countries, balance scales have been supplied to clinics and health centres by the United Nations Children's Fund (UNICEF). At boarding schools a good scale is often available in the kitchen, where it is used for weighing sacks of food. Similarly, in a village the local market master or the owner of a small shop will usually have a produce scale that can be borrowed. Special baby-weighing scales are necessary for accurate weight measurement of children under two years of age.

Height

Height is also a very important measurement in the assessment of nutritional status. As with weight, its interpretation in children is dependent on knowing the age of the child. Height should be measured with the subject barefoot. Though many different types of equipment are available, height can be fairly accurately measured with a tape-measure or a ruler. The following method may be used.

Locate a vertical wall rising from a truly horizontal floor. Make a horizontal pencil line about 2 cm in length at a height of 1 m from the floor (60 cm for children). Then, using sticking plaster, sticky tape or a drawing-pin, secure the bottom of a 1-m length of tape-measure to correspond with the line. Similarly fasten the top, which will now be 2 m from the floor. The person being measured stands against the wall facing outward. The height of the individual is ascertained using a block of wood having a true right angle. A rectangular block with the dimensions 30 x 10 x 20 cm is adequate, although a triangular block of the same dimensions is easier to handle.

The measurement of the length of young children presents more difficulty. A suitable apparatus consists of a flat board of dimensions 120 x 40 x 2 cm with a headboard 30 cm high fixed at a right angle to one end. The triangle used for height measurements can be used as a sliding foot-piece. A metal tape-measure is nailed to the board for readings in centimetres.

A less satisfactory alternative is to push a flat wooden bench, available in most dispensaries and schools, up against a wall in the corner of the room and measure it off in centimetres, starting about 50 cm from the wall and going up to 150 cm. The triangle is again used as a foot-piece.

When the length of an infant or toddler is being measured, the child must lie flat and straightened out to full length. For research purposes or where adequate funds are available, commercially made length boards can be used.

Series Readings

A series of readings of weight and/or height of an individual taken at, for example, monthly intervals gives valuable information. In an adult, weight loss indicates that energy intake is below energy output. Gain in weight indicates a more than sufficient energy supply. In adults a series of weight readings might be used, for example, during a famine to ascertain whether relief measures are adequate, or in a normal year to see if weight drop occurs during the hungry season. In children a series of monthly height and weight readings gives an extremely valuable record of the child's progress and nutritional status. It is worthwhile to keep a record of measurements taken of the heights and weights of children in schools, dispensaries and even community centres. The measurements can be carried out by either medical or non-medical personnel. When weight is measured in a series of readings, the figures are useful even without those for height.

If single readings of weight or height are available, they can be compared with a standard weight or height. The individual child's actual weight or height can then be expressed as a percentage of that expected for his or her age or in terms of standard deviations or Z scores.

Weight for Height

When weight and height have both been measured, it is possible to determine how near the child is to the standard weight for height. Even if the age of the child is not known, it is possible to assess nutritional status to some degree by expressing the weight as a percentage of that expected for the child's height or length or in terms of standard deviations or Z scores. This figure gives a relative measure of how thin the child is. Another commonly used method is to calculate the body mass index (BMI).

Mid-upper-arm circumference (MUAC)

The measurement of the circumference of the left upper arm midway between the acromion process (the bony tip) of the shoulder and the olecranon process (the point) of the elbow is being increasingly used as an index of nutritional status. Fibreglass tape-measures that do not stretch should be used. This method does not provide nutritional status information as reliably as does measurement of weight and height, but it has the advantages of being inexpensive and usable where no scale is available for weighing. Furthermore, between about eight months and five years of age the standard arm circumference increases very little. An arm circumference above 13.5 cm can be considered normal for children from one to five years of age. MUAC between 12 and 13.5 cm indicates moderate malnutrition, and below 12 cm indicates more serious malnutrition. The MUAC measurement may be especially suitable for use by persons with a minimum of training or for gross assessment of nutritional status in famine areas.

Head and Chest Circumference

The head circumference can be measured using the same tape-measure used for MUAC. The tape is placed horizontally around the head at a level just above the eyebrows, the ears and the most prominent bulge at the back of the head. Head circumference is related to brain size, but brain size is not necessarily related to intelligence.

The chest circumference is measured horizontally at the nipple line. Up to six months of age the head circumference is usually larger than the chest circumference. Children over 12 months of age having a head circumference larger than the chest circumference are abnormal; this is evidence of poor growth of the chest.

Skinfold Thickness

The skinfold thickness can only be measured if a pair of skinfold callipers is available. This instrument is designed to measure the thickness of the skin and subcutaneous fat using constant pressure applied over a known area. The two most common sites for measurement are over the triceps and in the subscapular region. The measurement is of considerable value in assessing the amount of fat and therefore the reserve of energy in the body. Unfortunately this instrument is rarely available in small hospitals, let alone health centres and dispensaries. This situation could easily be rectified, since the instrument is not expensive. The two most common skin callipers used are the Harpenden, made in the United Kingdom, and the Lange calliper, made in the United States.

LABORATORY TESTS

Many laboratory tests have great value in determining nutritional status, but few of them can at present be performed outside large hospitals. Only those tests that are widely available are discussed here.

Haemoglobin. An accurate assessment of haemoglobin level is by far the most important laboratory information that can be obtained in any nutrition survey. Accurate haemoglobinometers are rarely available in district hospitals, health centres and dispensaries. However, some cheap and simple-to-use haemoglobinometers which are reasonably accurate are now available.

In hospitals and for field research the cyanmethaemoglobin method is recommended. Blood is collected from a finger, ear lobe or heel prick. Two measured samples of 0.02 ml of blood are added to Drabkin's solution (a cyanide-ferricyanide solution). The specimen should be stored cool and protected from sunlight. The haemoglobin is determined later the same day using a spectrophotometer or other apparatus.

Haematocrit or packed cell volume (PCV). This determination is also important in the diagnosis of anaemia. A capillary tube is filled with blood from either a vein or finger prick. The sample is spun in a standard electric or hand centrifuge, which separates the red cells from the plasma. The haematocrit or PCV is the percentage of the blood volume composed of red cells.

Red cell counts and blood films. Red cell counts are not easy to do and add little information to the above tests. However, it is easy to prepare a thin blood film on a glass slide. Such slides are useful, since they enable the size and uniformity of the red blood cells to be seen. Use of such slides may facilitate the diagnosis of malaria and the haemoglobinopathies that may also cause anaemia.

Serum protein. Determination of total serum protein and especially of the serum albumin and globulin levels can only be undertaken in a well equipped laboratory. These data are useful in cases of kwashiorkor, but they have not been found helpful in the diagnosis of mild or moderate PEM.

Examination of stools, urine and blood for parasites. After haemoglobin estimation, the next most important laboratory tests in a nutrition survey are strictly non-nutritional. There is little doubt that parasitic infestation and malnutrition are closely linked. The medical nutritionist must examine the individual and the community on all aspects related to public health. Laboratory examination should therefore be made of stools for the ova of hookworm, roundworm, Trichuris species,Schistosoma mansoni and other parasites; of urine for albumin, casts and Schistosoma haematobium; and of blood for malaria parasites. These tests are all easily performed in most dispensaries. They require only a microscope, a hand centrifuge, some laboratory glassware and a few simple reagents. Precautions should be taken in collection and disposal of specimens. Quantitative tests to assess the parasite load should be performed if possible.

During a nutrition survey it may be preferable to do these examinations on a separate day or during the afternoon following clinical examinations in the morning. In a large community it is advantageous to restrict these examinations to one particular group, such as all the children at the local school. The results will give a reasonable picture of the prevalence of diseases such as malaria and hookworm in the community. It is easier and more hygienic (especially with regard to stool examinations) to deal with a selected group than to collect specimens from people scattered over wide areas who have assembled in large numbers at a centre for clinical examination.

Biochemical tests. Certain biochemical tests are useful for assessing deficiencies of almost all the minerals and vitamins. Even though in many developing countries vitamin A deficiency and IDD are important public health problems, very few local hospitals have laboratories that can conduct tests to assess these deficiencies. Similarly, in countries where pellagra, ariboflavinosis and rickets occur there are very few laboratories that can assess these deficiencies.

Dietary Surveys

Accurate assessment of the dietary intake of a community takes much longer than getting a picture of the community's nutritional status by clinical or anthropometric examination. There are two main types of dietary survey. One relies on direct observation of a sample of the population, with their food measured and weighed over a given period of time. The other relies on inquiry, with a larger group of people questioned about their diet. Each type has a disadvantage: the former is very time consuming, and the latter depends on the memory, integrity and intelligence of the subjects questioned. Neither method takes account of past consumption or of uncertainties of food composition. Such involved methods are rarely justified or practical. It is often better to use cruder, simpler methods that provide data that reveal the causes of malnutrition and suggest corrective measures. The various methods of dietary survey are discussed below.

Observation. The only way to assess the diet accurately is to weigh and measure all the food that individuals eat over a representative period of time. A survey team goes to households and weighs and measures all food that is prepared, cooked and eaten, as well as that which is wasted or discarded.

If possible, the proportion of the total quantity of food prepared that is eaten by each individual should be weighed. (This is difficult in countries where household members often feed from one large communal dish or pot.) When the food eaten by each person on an average day has been ascertained, it is necessary to calculate the amount of each nutrient eaten by each subject or each family, using quantitative tables of dietary constituents.

A dietary survey of this kind requires a survey team of at least two persons that can cover two to four families at one time and perhaps 20 families in a month. It is essential to obtain truly representative households as samples and to cover a small, statistically acceptable sample of the population properly, rather than to try to cover more families in a less thorough manner.

Inquiry or recall. Direct inquiry cannot give very accurate information on amounts of energy or nutrients consumed. However, it can give an indication of the frequency of meals and the methods of food preparation and cooking, as well as providing details of the foods commonly consumed.

In developing countries it is most usual for a survey worker to go to a household and ask questions of the wife of the head of household. The answers are recorded on a form. This kind of inquiry depends heavily on the memory of those giving information and also on their attitude towards the person inquiring. False answers are often given unconsciously, or the subject may have some concealed reason for misleading the inquirer. For example, if the subjects believe that the inquiries are being made to ascertain whether famine relief food should be issued or increased, then quite naturally they will indicate that they are eating a small quantity and variety of food. If, however, they believe that the questioner is attempting to assess their standard of living or their degree of development, local pride may influence them to overstate the quantity and variety of food that they eat.

The most common method is to ask the subject to recall what was consumed during the previous 24-hour period. This is termed the 24-hour recall method. It is useful to have available local measures (bowls, cups, spoons) so that the respondent can indicate the approximate amount eaten.

Another survey method is to have literate people fill in a questionnaire. For example, schoolchildren may be given a questionnaire on which they are asked to record each morning for a week what they ate during the previous 24 hours. The process should be repeated at different seasons of the year. Such an inquiry gives no indication of quantities consumed, but it may provide useful information about meal patterns, the staple foods of each household, the frequency of consumption of certain foods such as meat, fish, eggs, fruit or vegetables, seasonal variations in diets, etc. Food frequency surveys of this kind can be performed on other groups of people. They provide qualitative, not quantitative, information.

Combined observation and inquiry. In a combination method, the observer goes to previously selected households and asks the wife of the head of household to show what food she intends to cook for the family that day. This food is then accurately weighed. The worker also records the number, sex and age of the people in the household. He or she then moves on to the next household. Clearly much more ground can be covered per day using this method than with a full-scale dietary survey as described above.

However, the respondent may have no idea of how much food she is going to use that day, or she may exaggerate the amount. This type of survey takes no account of food loss or wastage and gives no indication of what individual members of the family consume. The medical nutritionist is often very keen to know what the toddler or the pregnant woman actually eats, not what is prepared for the whole family.

One survey using this method, carried out in East Africa under the direction of statisticians, reported that the people surveyed consumed over 5 000 kcal per head per day. Malnutrition and undernutrition were known to exist in this area, and the likely intake of those questioned was 2 200 kcal. Clearly the average householders in this survey area had tried to impress the observer with how well they were living.

Reducing random and systematic errors. In almost all methods of obtaining dietary information there are common errors which make the data unreliable or even lead to wrong conclusions. These errors can be random or systematic. Various precautions including quality control can be taken to reduce some errors. No dietary assessment measurements are completely precise.

Random errors are related to the precision of the dietary method used. If the number of observations made is increased, the influence of the random errors on conclusions reached will be reduced. Many such errors cancel each other out, and they are therefore of less concern than systematic errors.

Systematic errors cannot be reduced by increasing the numbers of observations, and they do not usually cancel each other out. They are often cumulative and may be increased when more observations are taken. They therefore constitute a more worrying problem than random errors.

Systematic errors may result from several kinds of bias. Possible biases on the part of the interviewer include improper writing down of answers; neglecting to ask certain questions; and failure to ensure that the subject understands the questions. Those on the part of the subject include provision of information that is not true but is believed to be the "desired" answer (perhaps to try to create an appearance of being either better off or worse off); underreporting or overreporting of the consumption of certain foods; and lack of understanding of certain questions.

Other major sources of error in dietary surveys include difficulties in estimating the size of food helpings or the size of an item eaten; poor memory of what foods were eaten; and failure to remember or to mention foods eaten between meals. Errors may also arise in translating the results recorded on the survey form into amounts of food in grams and millilitres and into nutrients consumed There may also be coding errors.

Methods that should be used to try to minimize errors include quality control; training, retraining and checking of interviewers, coders and data analysts; use of standard questioning methods and good data collection forms; consistent use of good and appropriate food models of different sizes and commonly used household measures and utensils; and finally instilling into survey workers and study subjects the vital importance of accurate information. Interviewers should understand that it is much better to admit errors rather than to hide them or falsify data. Respondents should be convinced that it is preferable to admit not knowing or not remembering rather than to provide an untrue answer.

Vital Statistics

Vital statistics are those related to births and deaths in the community. Complete and accurate vital statistics are not maintained in all countries, nor are they likely to be in the near future. However, vital statistics are so important as an index of nutritional status and for other public health reasons that they serve a useful purpose even if collected in small areas only. Infant mortality rate (death during the first year of life) gives a good indication of the state of nutrition and the health of the community. The neonatal mortality rate (death during the first month of life) and stillbirth rate are also useful.

In developing countries, figures for the toddler mortality rate (TMR) (deaths between the first and fifth birthdays) are far more useful to the nutritionist than the other rates TMR values often give a good indication of the prevalence of PEM, although they do not necessarily illustrate the nutritional status of the whole community.

TMR often provides a clear indication of the comparative state of development of a country. For example, in Scandinavia, the former Soviet Union, North America and the United Kingdom, TMR is below 1 per 1000, while in much of Asia and Africa it is at least 35 times as much. The infant mortality rate is around 7 per 1000 in Sweden and from 35 to 150 per 1000 in most African countries.

Although it is normally impossible for an individual worker or a survey team to collect accurate vital statistics, some information of value regarding birth rate and death rate is usually available. For example, during a survey, one can easily ask all married women of child-bearing age two simple questions:

— To how many live children have you given birth?

— How many are still alive today?

From these figures a percentage of children that have died and also some indication of the fertility rate can be obtained. Careful questioning might also elicit the approximate ages of the living children and a rough estimate of the age at which the others died. Questioning as to the cause of death, if carefully done, may produce useful information.

It must be emphasized that information gathered in this way provides only rough estimates of the true figures, but these are nevertheless useful and will have to suffice until such time as proper vital statistics are maintained

Other Useful Data

Many other types of information are helpful in assessing nutritional status. These include other health statistics and medical information. Diarrhoea rates, measles incidence and other disease data have implications for nutritional status.

Since food security is partly dependent on food production, agricultural data are useful in judging the likelihood of food security and its relationship to nutrition. Economic data provide information for judging the nutritional climate in a community or a country. Figures on incomes, purchasing power, food prices and food distribution are useful. Data normally obtained by food scientists are helpful in judging nutritional status, food quality and food safety.

Participatory and Rapid Appraisal Techniques

In the field of nutrition, as in social, agricultural and other fields, it has been increasingly realized that participatory methods of collecting information have many advantages. Involving members of the community, the potential beneficiaries, at the stage of data collection can prove extremely valuable. The active participation of the community in assessment and analysis, rather than only in the action stage of a project or activity, is likely to be very helpful. It assists in educating the public, in mobilizing local resources, in empowering people and in sustaining the success of actions taken. The community members, whether villagers or urban dwellers, come to understand their health and nutrition situation and the underlying causes of various problems. They offer alternative options for change and play the central part in implementing actions. This kind of participatory development, which is now suggested for nutrition, was well described 30 years ago by Paulo Freire working in Brazil. He termed it "conscientization" of the community, or helping community members become more aware of the causes and consequences of nutritional problems and, more important, how they can work together to prevent and overcome such problems.

A new series of techniques have emerged in the last decade as tools for participatory appraisal exercises. Semi-structured interviews, with either selected individuals or focus groups, are combined with observation (e.g. transect walks) and visualization techniques (such as mapping, seasonal calendars, ranking exercises, time charts and Venn diagrams). These techniques are particularly useful to gain an understanding of people's food habits and related beliefs, food entitlements and existing constraints and the role of the different family members in relation to nutrition (household food security, health and care). The choice of the techniques and their combination will be determined by the information needs and time constraints of community members. It is essential to cross-check the information gathered through different techniques. The information must be analysed on a regular basis to identify inconsistencies and remaining gaps, to be addressed in the next stage of the appraisal.

Participatory appraisal can best be carried out jointly by the community and local development staff, as it is a continuous process and should be an integral part of development activities at community level (for identification and selection of activities to promote household food security and nutrition, monitoring and evaluation, and reformulation).

Another major change in data gathering for assessing the nutrition situation of communities is the acceptance of rapid appraisal methods. Rapid appraisal exercises can help develop a first understanding of the situation and identify issues on which further information is needed. They can then be complemented by formal surveys or routine data collection. The rapid methods borrow from anthropology and the other social sciences to obtain both quantitative and qualitative data. They offer promise because if properly used, they can provide useful information without the need for more complex survey methods or very large sample sizes. Even though rapid appraisal is usually carried out by international or national experts, it should involve local development staff who will be in a position to ensure follow-up of the process within their regular activities.

Nutritional Surveillance

Nutritional surveillance is a set of activities to assemble information to assist in policy and programme decisions to influence the nutritional status of a population. It usually includes the regular and timely collection, analysis and reporting of nutrition-relevant data. Surveillance differs from surveys in that it involves the periodic or continuous collection of data.

For many years various kinds of nutritional information have been collected, often for decision-making, but nutritional surveillance did not become a central activity in national nutrition planning until after 1976, following the report of a Joint FAO/UNICEF/WHO Expert Committee entitled Methodology of nutritional surveillance.

Because nutritional status is influenced by many different factors, nutrition monitoring and nutrition indicators may come from many different disciplines and may be of many different kinds, ranging from meteorological data, to food production, to nutritional status of people and so on.

Because nutrition is an outcome of social, economic, health, agricultural and other conditions, the nutritional status of a population can be used as an indicator of the overall development of a society. Specific nutritional status indicators are often better indicators of equitable development than are traditional economic indicators such as gross national product.

Information for Decision-making

Nutritional surveillance, like nutrition surveys, is not useful if the data collected are not used to improve the nutritional status of the population. The weakest part of many nutrition surveillance programmes has been that the data collected have not been used to solve nutrition problems. For various reasons decision-makers have not used the information to take action. Why? It may be that lack of information was not the problem, that the kind of information needed is not being provided, or that there is a lack of commitment and resources to solve the nutritional problems. In general, it is agreed that the information needs to be provided in an easily understood form and in a timely manner.

In the past, nutritionists, health workers and others collected data and passed them to decision-makers in the expectation that action would follow. Some rethinking is strongly recommended. It is suggested that the first step after identifying the important nutritional issues should be to discuss and review possible policies and programmes and to identify how decisions will be made to influence these policies and programmes. This exercise would influence decision-makers to identify for themselves the information that they need in order to make decisions. If this approach were taken, the data collected would be what the decision-makers needed and would be likely to be used by them. The data would be analysed and would be discussed with the decision-makers, and decisions could be made to take appropriate actions. Later the impact of the actions would be determined.

Before surveillance is initiated, there should be an assurance first that there will be good communication between the people and institutions collecting the data and second that the data will reach the people and institutions that have the power to make decisions.

Assessing and Monitoring Nutritional Problems

There are a huge number of possible indicators of nutritional status. The following are some typical indicators of different kinds that have been used in nutrition monitoring.

— food crises:
 - production patterns,
 - market prices,
 - food stocks,
 - fall in body weights;

— protein-energy malnutrition:
 - children's anthropometry
 - children's growth,
 - infectious disease rates,
 - food intake relative to need,
 - body mass index;

— household food security:
 - employment levels,
 - market prices,
 - changes in real income and purchasing power,
 - dietary energy supply;

— caring capacity:
 - maternal education,
 - literacy rates,
 - maternal employment,
 - public expenditure,
 - breastfeeding (duration and percentage);

— malnutrition-infection complex:
 - incidence of diarrhoea,
 - immunization coverage,
 - availability of clean water,
 - children's weight for age;

— micronutrient deficiencies:
  - iron deficiency: rates of anaemia,
  - vitamin A deficiency: night blindness/xerophthalmia in children,
  - iodine deficiency: goitre, cretinism;
— non-communicable chronic diseases:
  - cardiovascular disease, diabetes, obesity, some cancers:
  - age distribution of population,
  - age-specific mortality,
  - changing dietary and lifestyle patterns.

Local decisions need to be made on which indicators to use. It is best if only a few indicators are chosen and if these are suitable for relatively easy regular collection. In developing countries the most widely reported indicator of malnutrition is low weight for age. However, data used are often not representative of the population and have been gathered from hospitals or growth monitoring clinics. For nutritional surveillance the data should be representative of the targeted population (for example, children six to 36 months of age of a particular district) and should be collected periodically. The use of well-chosen sentinel sites where data are regularly collected is a means of obtaining such data. However, although weight-forage data provide a picture of the nutritional status and, if collected regularly, give important information on trends, they do not reveal the causes of the malnutrition identified. These underlying determinants can be grouped into those related to food security, health factors and child care. Data are often collected routinely on some of these causes.

In food crises early warning indicators may allow action before overt starvation. Indicators may be based on forecasts of food availability and food prices in the market. In countries where droughts are common, data on rainfall provide an early warning; these data are followed by food crop status and harvest yield estimates plus monitoring of food stocks, reserves, marketing and prices. Sentinel households can provide useful information, some quantitative (e.g. crop yields and food stores) and some more qualitative (e.g. subjective views about family food security and reporting when they have to sell their personal possessions to purchase food).

In relating health factors to nutrition the focus is usually on infections and on monitoring infectious diseases such as measles, whooping cough, diarrhoea, respiratory infections, intestinal worm infections and malaria. Important health interventions also deserve monitoring; these include immunizations, oral rehydration for diarrhoea, attendance at clinics and preventive measures such as health and nutrition education, sanitation and improvement of water supplies.

To monitor caring practices and their impact on nutrition, data could be collected on breastfeeding and weaning, time available to the mother for child care and competing activities, differential treatment of girls and boys, family responses to poor appetite or poor health in their children, etc.

Many of the indicators discussed above are rather directly related to PEM, but many are also associated with micronutrient deficiencies. Lack of food security, high rates of disease and deficient caring practices have a negative impact on vitamin A and iron nutritional status as well as on PEM. Specific micronutrient deficiencies may also be monitored, for example by monitoring night blindness rates for vitamin A deficiency or haemoglobin levels for iron deficiency. Objective data might be collected from sentinel households. Data on food consumption also provide useful information.

The use of rapid appraisal methods is potentially valuable in monitoring nutrition. Some of the data collected in this way might be qualitative, including some that provides information on the functioning of relevant programmes.

Nutritional Surveillance Systems

There are four types of nutritional surveillance, distinguished by their different objectives. A number of countries have only one type of surveillance system, while others have several or even all four. Where several types are used, they may be coordinated in an organized way and may use some common data.

Timely warning and intervention. Nutritional surveillance was first established to warn governments of poor nations of imminent nutritional crises. It was in part modelled on health surveillance for important infectious diseases. Certain infectious diseases such as plague and cholera are notifiable to WHO; countries require that each district or province notify the national ministry of health on a weekly basis of the number of cases of notifiable diseases. In famines or severe crises, data on famine deaths or serious famine-related malnutrition can be collected and reported. Unlike outbreaks of serious infectious diseases, famine brings with it many cases of serious malnutrition. Nutritional surveillance reports on indicators that would warn a government of an approaching nutrition disaster. As listed above, production patterns, market prices, food stocks and fall in body weights are possible indicators of food crises.

The types of data needed for an early warning system must be decided in the individual country or the affected region of the country. They cannot just be prescribed. It is important that the indicator system be sensitive and that it be able to predict food crises, even if warning is sometimes given of a crisis that does not then occur.

The first indicator may be rainfall below a certain level over a period of two or three agriculturally critical months. The next set of indicators might relate to the important crops in the field prior to harvest. These may be followed by estimates of food production and indications of food consumption. Finally, actual indicators of nutritional status such as the weight of adults and children in poor families may be monitored.

In some countries indirect indicators have proved useful, such as the pawning of household items, the movement from the consumption of a preferred food such as rice to a less desirable food such as cassava or the actual measure of food stores of sentinel households.

In Indonesia a timely warning intervention was introduced at the district level in drought-prone districts. Data collected at the district level could be provided quickly to the district official, who was given authority to take immediate action. A district-level food security system was established so that surveillance data indicating a shortfall in the food supply would result in delivery of a supply of rice to the local markets to prevent price rises and unavailability of rice. If the data had had to go to the capital city for review before decisions were taken, as is the case in many countries, long delays would have occurred. This example illustrates the need for data to be provided rapidly to officials authorized to take action speedily. Unfortunately the need is not often satisfied; data often end up as reports considered by people far from the scene, on which little action is taken.

Nutrition surveillance for policy and programme planning. Many kinds of indicators, including those listed above, can be used by governments or local authorities for surveillance to influence policy and programme planning. The data may be on nutritional status or on a variety of factors that influence nutrition. For example, anthropometric data may be collected on a regular basis to describe PEM trends over time. The data may be analysed to discern groups of the population most severely affected. They might be used to show which five provinces in a country have the worst malnutrition; which social groups are worst off; or what health factors are related to the most serious PEM. The next step might be to decide on direct interventions (perhaps supplementary feeding or nutrition education) for the most seriously affected groups and to suggest ways in which existing policies (for example, regarding credit for small farmers to improve agricultural productivity or subsidies for staple foods for the poor in urban areas) might be modified or strengthened to influence nutritional status.

Costa Rica has had a national nutritional surveillance and information system since 1978. The system is designed to target activities to the poorest parts of the population and the poorest areas of the country. The anthropometric data used include child height, collected when children enter primary school, and the weight of younger children, collected by home visiting. A goal of the surveillance has been to use existing programmes more effectively by targeting activities to the poorest families who have the most PEM.

In these types of programmes interventions may be clearly nutritional (supplementary feeding; iron supplements) or non-nutritional but expected to have an impact on nutritional status (measles immunization; improved sanitation and water supplies; actions to reduce women's work load).

Nutritional surveillance for management and evaluation. Surveillance can be used to evaluate programmes aimed at improving nutrition and to assist in their management. For example, data from growth monitoring over a period of five years might be used to evaluate whether an agricultural credit scheme has improved the nutritional status of children; or night blindness data might be used over time to evaluate whether horticultural activities are influencing vitamin A nutritional status.

Data collected might be used as an internal management tool to judge the efficiency with which programmes in different parts of a country reach their objectives, or to compare the effectiveness of two alternative interventions aimed at solving the same nutritional problem.

Nutritional surveillance for advocacy. Scientists are often reluctant to be advocates, believing falsely that advocacy is unscientific. It is highly desirable, however, that most of those involved in nutrition be advocates for action. If serious problems of malnutrition are found in areas where food and health services are available the situation is unacceptable, and it is right to advocate interventions to reduce malnutrition.

Conducting surveillance for advocacy mainly involves collecting data on the prevalence of PEM or micronutrient deficiencies or on related indicators and using these data to get support for action. Support can be solicited in different ways including making the government aware of the problems found or embarrassing the government into taking action by publicizing a serious nutrition problem in the news media. The objective is to influence policy-makers to allocate resources and to provide the needed assistance to allow interventions or programmes to be implemented to improve the nutritional status of the communities affected. For example, in Chile it appeared that a reduction in supplementary foods provided to poor families was adversely influencing nutritional status. Advocates used anthropometric data from the health monitoring system which showed a recent rise in the rates of malnutrition in children. When the government was presented with these findings, it reinstituted the supplemental food benefits.

References

Cannon, G.C.1992. *Food and health: the experts agree.* London, UK, Consumers' Association.

WHO.1976. *Methodology of nutritional surveillance.* Report of a Joint FAO/UNICEF/ WHO Expert Committee. WHO Technical Report Series No. 593. Geneva, Switzerland.

FAO/WHO. 1992b. *International Conference on Nutrition. Nutrition and development a global assessment.* Rome.

Gibson, R.S. 1990. *Principles of nutritional assessment.* Oxford, UK, Oxford University Press.

World Health Organization (WHO). 1966. *The assessment of the nutritional status of the community, by* D.B. Jelliffe. WHO Monograph Series No. 53. Geneva, Switzerland.

6

Food Quality and Safety

Food production and food demand receive a great deal of attention in agriculture and nut families have to have access to sufficient food which then is consumed in adequate amounts by each family member. What receives less attention in writing, in training and in action is the fact that the food and water that people consume need not only to be adequate in amount, but also to be safe and of good quality.

Most industrialized countries have well-developed systems to ensure a reasonable level of safety and quality of foods consumed. Most developing countries have rudimentary systems that need strengthening. For a food system to work effectively, all those involved in it - from production, through processing, to marketing and eventual consumption - must be educated about food safety and quality and must implement actions to ensure them. Consumer education is a part of this effort.

Consumers, the food industry, government ministries and international agencies all have important, interrelated roles in ensuring food quality and safety. Food control measures can help reduce food losses and food spoilage, promote appropriate food processing and help ensure food safety and quality for the local consumer, for local markets and for export.

These lofty goals require appropriate legislation, regulations and food standards. These in turn demand a means to ensure compliance, which entails surveillance or monitoring, usually carried out through food inspection and in many cases laboratory analyses. Poor countries may not have the trained staff or the facilities to do a very good job in this respect, so they often decide to limit their activities in the area of food safety by trying to avoid serious outbreaks of food-borne illnesses and serious food contamination. Without much laboratory backup, public health inspectors and related staff may visually inspect meat at abattoirs or in meat markets; visit shops to find spoiled food; and inspect restaurants, hotels and stalls that sell food. They can insist that reasonable standards of hygiene are observed.

Appropriate national authorities should at the very least take steps to educate the public about food safety and quality so that consumers can insist on safer, better foods. These practices may begin with the education of the farmers who grow the foods and continue with education at various stages along the food path, up to the family kitchens in rural and urban areas. Education and assistance to food processors and manufacturers are also important. All should be made aware of standards, food laws and existing regulations and how to adhere to them.

In many rapidly urbanizing poor nations, more and more food is sold, processed, cooked and even served by small-scale entrepreneurs such as market or street-corner vendors. At street-side stalls or tables food safety and quality practices are not infrequently ignored. As many students are frequent street-stall users, food safety and food quality should be included in nutrition education activities and in the curriculum in schools to empower students to recognize food that is of doubtful quality or safety.

Ensuring Food Quality

Poor countries often do not have the institutions or the personnel to ensure food safety and control, although most do have some legislation, standards and regulations on the books. Governments would be wise to seek help internationally to improve their capacity in this regard. Small, poor countries can sometimes, with international assistance, share food microbiology and toxicology laboratories. The larger, emerging developing countries, sometimes called middle-income countries, should devote much more effort than they do now to ensuring food safety, and many can afford to do so. These countries are becoming highly urbanized and commercialized. The centres of the cities often appear modern and Western, with high-rise buildings, paved streets and running water in every household. However, nearby are often slums and squatter settlements which do not have safe water or satisfactory sanitation. In these areas the food that is sold is very likely to be contaminated and unsafe.

The food industry has an important part to play in food quality and safety at every stage of the food path from agricultural production onwards. For example, in the fields where crops are grown chemical fertilizers and pesticides need to be properly used; appropriate methods of preservation and storage need to be adhered to; and good technologies must be adopted to ensure food products at low cost but of high quality and safety.

International organizations can provide expert technical assistance and advice on various aspects of food quality and food safety, including the use and control of food additives; cut-off points for acceptability of food contaminants; and monitoring of simple hygienic practices in particular industries.

FAO and other organizations have a very important role internationally in advising member countries about appropriate legislation and regulations, which may include specific standards and guidelines related to quality, safety and labelling of foods put up for sale. Many of the standards and guidelines have been developed by the Codex Alimentarius Commission, a joint body of FAO and the World Health Organization (WHO) that provides international standards designed mainly to protect the health and welfare of the public while ensuring fair trade practices. These food standards help in international trade of food products. FAO, almost since its inception 50 years ago, has helped member countries improve the quality and safety of foods available for consumption by their people, through its staff expertise, meetings and expert consultations, numerous publications, assistance in standards development and numerous other activities. But for the countries themselves, adhering to standards and codes that help ensure the safety of foods must surely be considered a part of national or local food security.

An epidemic of a serious food-borne disease can have a significant negative impact on food trade within a country or internationally. A good recent example is the cholera epidemic which was first reported in Peru in 1991 and then spread first to other Andean countries and then to a wide group of Latin American and Caribbean countries. Peru is a major exporter of seafood, and very soon its trade was greatly affected, areas were quarantined and internal trade was limited. The result was a major negative impact on many poor people involved in the trade of seafood and later of other foods as regulations became extended. The epidemic led Peru to pay much greater attention to urban water supplies, sanitation, food handling and sale of street foods.

Food or water introduces health risks if it is contaminated with pathogenic organisms, toxins, pesticides or poisons. Any of these can lead to illness, sometimes in a few hours and sometimes a long time after their consumption. Perhaps the most common symptom or sign of illness resulting from consumption of contaminated food is diarrhoea. Diarrhoea can be caused by viruses, bacteria, parasites, toxins or poisons. An example of disease occurring a long time after the consumption of contaminated food or water is the development of certain cancers because of carcinogenic toxins.

Contaminated foods eaten at home or in public eating places may appear to be safe or may have evidence of contamination. If food, beverages, dishes or utensils are obviously unclean, if the food looks or smells bad, if a food that is meant to be eaten hot is served cool or lukewarm, if the environment where the food is served has flies, cockroaches or evidence of rodents or if food servers have dirty hands and clothes, then it is likely that the food being served is contaminated.

Sometimes it is difficult for people to refuse to eat food that they suspect may be contaminated. However, there are some steps that consumers can take, for example, at a stall selling street foods.

— Choose a food that is served at a very high temperature. If it is cold, ask the vendor to heat it more. Heat kills many organisms.

— Of uncooked foods, choose only those that are eaten peeled. Choose a banana rather than a slice of watermelon, for example. Even if both are crawling with flies, the banana can be peeled and then kept free of flies.

— Ask for a drink that comes in a bottle or can that can be opened by the consumer, or order tea or coffee that is served very hot.

Remember the old saying, "If you can't boil it, bake it or peel it, then forget it!" This saying makes a lot of sense.

Steps to Improve Food Safety

In every household, but especially in those with less than ideal sanitation, some knowledge about food-borne disease is very important. It should be imparted in every school and should be an element of health education at every level. Many people in developing countries have very little understanding of the germ concept of disease, i.e. that serious illness can be caused by unseen organisms. An important challenge for health educators is to ensure that people understand that microorganisms cause disease.

Diarrhoea is commonly caused by a variety of microorganisms which are in human faeces and get into food and water. The following simple preventive steps can be taken.

Latrine and Faeces Disposal

The first sanitary essential in the household is a latrine and a well-organized system of safe disposal of human excreta. Measures are needed to prevent human faeces from contaminating the household and its environs. Very young children may not be able to use a pit latrine, but their faeces can spread disease and therefore need to be disposed of safely. Animal faeces are not nearly as dangerous as human faeces, but they can spread disease.

Personal Hygiene

All members of the household should understand the basic rules and practices of good personal hygiene and practise them. Hands should be washed after use of the latrine and before each meal, and by people preparing food. All aspects of personal hygiene, however, including a clean body and clean clothes, also have a role. Personal hygiene is much easier if adequate water is available.

Household Hygiene

A third form of protection is to ensure a good level of household hygiene, which is especially important in the kitchen and wherever food is stored, prepared and eaten. These places need to be kept clean and as free as possible of vermin such as flies, cockroaches and rodents. A clean house is a protection against food contamination and resulting disease.

Food Preparation and Storage

In the home, irrespective of its circumstances, the best possible efforts should be made to store, prepare and serve food in a way that minimizes the dangers of contamination and to make the meals as nutritious and as appealing as possible. This is relatively easy for an affluent household that has a refrigerator, a gas stove, running hot and cold water in the kitchen and a flush toilet. For a poorer household where there is no refrigerator, where food is cooked outside over a wood fire, where water is carried for two hours from a contaminated stream and where there is a pit latrine, food hygiene is a struggle.

Food Preparation to Ensure Food Safety

Bacteria that cause disease multiply rapidly in many foods, but more rapidly in foods of animal origin that are warm and wet. Small amounts of sugar enhance bacterial breeding, while larger amounts reduce it. If foods are not stored at low temperatures millions of bacteria will breed in them. Meat stew will deteriorate very quickly, stiff maize porridge moderately quickly and bread less quickly.

Uncooked dry rice granules will not deteriorate quickly. It should be understood that parasitic eggs (such as those of roundworm) or parasitic cysts do not multiply in food, but they do cause disease.

Although many cooked foods in a poor household without a refrigerator should not be stored for very long, it is helpful to cover food, perhaps with gauze, to let in air but not flies. Alternatively, food can be kept in a simple "meat safe" which can be a simple wooden box on legs with metal or plastic screening on the sides or across the front. Each leg of the meat safe can be stood in a bowl or can of water to prevent ants and cockroaches from entering the safe.

Biological Contamination of Food

Organisms are much more common contaminants of food and causes of disease than toxins or chemical poisons. More than 25 organisms, including bacteria, viruses and parasites, infect humans and cause specific disease after being consumed in contaminated foods. Microorganisms are ubiquitous, but only some of them are pathogenic (disease-causing) in humans.

Many of the pathogenic microorganisms are passed out of the body in faeces; they infect another human being when they reach the mouth, taken there perhaps by unwashed hands, utensils or flies. This type of transmission is termed faecal-oral transfer.

Gastro-enteritis or diarrhoea resulting from toxins produced by microorganisms can be distinguished from diseases caused by microorganisms invading the lining cells of the gastro-intestinal tract. The spread of both types is similar. The most important types of microorganisms are listed below.

Viruses

It is now clear that many outbreaks of diarrhoea, particularly in children, are caused by virus infections, mainly rotavirus or Norwalk virus. These viruses do not multiply in food, but they do in the intestine. The measles virus can also cause diarrhoea.

Bacteria

Many different bacteria are food borne and cause gastro-enteritis and other diseases.

Many different types of salmonella have been identified and found to be pathogenic. In some countries salmonella is the main cause of food poisoning. It may be transmitted through consumption of raw or undercooked eggs or through contamination of foods with salmonella by food handlers. Usually symptoms begin less than 48 hours after the food has been consumed. The disease is self-limiting, usually ending within six days. Salmonella typhi leads to the serious disease called typhoid fever, which is also spread by faecal-oral transmission. It is characterized by intermittent fever, a rash, abdominal pain and great, sometimes lengthy, debilitation.

Some staphylococci, such as Staphylococcus aureus, a widespread organism, can lead to diarrhoea and vomiting. Clostridium (Clostridium perfringens orClostridium welchii) is a common cause of food poisoning. These bacteria are anaerobic and produce spores which can be widespread. Clostridium botulinumcauses a very virulent form of food poisoning. It is usually food borne, but it can also infect wounds. If consumed its toxin produces serious neurological and muscular signs and symptoms, and the disease is often fatal. In food-borne infections, the contaminated food, often a preserved meat, becomes the site for toxin production by the clostridia. The spores are resistant to heat, but the toxins are destroyed by thorough cooking.

The disease that used to be called bacillary dysentery is caused by four Shigella species that infect foods: S. sonnei, S. flexneri, S. dysenteriae and S. boydii.These bacteria lead to marked diarrhoea, sometimes accompanied by vomiting and blood in the stools.

A very serious bacterial infection is cholera, caused by the organism Vibrio cholerae. The infection involves much of the small intestine Cholera is an acute infection leading to profuse and frequent watery stools, vomiting and abdominal pain. The patient may soon become severely dehydrated and may die rapidly Oral rehydration can be life saving.

Other food-borne bacteria incriminated in diarrhoea or other diseases include certain serotypes of Escherichia coli (although many forms of E. coli are non-pathogenic); Campylobacter species; Bacillus aureus; and other vibrios such as Vibrio parahaemolyticus.

Parasites

Parasitic infections can be transmitted in food and water. The most prevalent intestinal worm infection is Ascaris lumbricoides (roundworm), which infects about 1 200 million people worldwide. Female worms in the intestine of an infected person produce millions of eggs which pass out in the faeces. If faeces are not properly disposed of, the eggs can get in the household environment or in dust being blown around, can get into food and can infect new subjects. Whipworm (Trichuris itrichiura) and the protozoan infection Giardia lamblia are spread in the same manner and can cause serious disease.

Other parasites are transmitted through consumption of raw or undercooked food. Pork or beef that is not thoroughly cooked may be infected with Taenia solium(pork tapeworm) or Taenia saginata (beef tapeworm) which if eaten will infect the consumer. Pork tapeworm is a particular danger, because it can cause cysticercosis with serious complications. Undercooked or raw freshwater fish may be infected with a tapeworm called Diphyllobathrium latum. The tapeworm in the human gut competes with the host for vitamin B_{12} so the infection may lead to macrocytic anaemia.

Non-infective Food Toxicity

Non-infective toxins or toxic substances in foods for human consumption can be "natural" in that they occur in nature. In certain fungi, lathyrus, cassava and fish, for example, these are the most common toxins. Less common but of great importance are toxins that are artificially added to food, such as various chemicals used to assist in food production, including fertilizers, weed killers, insecticides and fungicides. Other toxins that cause problems for humans include metals such as mercury or lead, which may get into the food supply or be consumed inadvertently.

Below is a summary of some of the more important non-infective substances that have resulted in ill health when consumed in food.

Aflatoxins

A toxin produced by a mould called Aspergillus flavus was found in 1960 to kill poultry fed groundnuts contaminated by this mould. A flurry of research followed, and it became clear that aspergillus grows on many foods, including cereal grains, when stored damp in tropical countries. In animals aflatoxin produces liver damage and carcinoma. It is not yet clear if aflatoxins are a determinant in primary carcinoma of the liver in humans; it now seems more likely that the high rates of primary liver cancer in Africa are a result of hepatitis earlier in life. Aflatoxin does cause disease, however. Some countries attempt to monitor the aflatoxin content of foods. Other hepatotoxins are found in food but they are not as important as aflatoxin.

Lathyrus

Lathyrus sativus is a vetch that grows wild, but it is also cultivated, particularly in India, where it may be planted in wheat fields. A neurotoxin in the plant, when consumed in large amounts, causes a neurological disease which can first result in weakness or spasticity in the legs and eventually lead to crippling and paralysis. The disease, lathyrism or neurolathyrism, has been widely discussed in Indian medical literature.

Fungal toxins

Some forms of fungi such as mushrooms are delicious foods and perfectly safe to eat. Other fungi, some of them resembling mushrooms, are highly toxic and lead to gastro-intestinal symptoms and perhaps kidney damage. Consumption of food contaminated with the fungus Claviceps purpura leads to the disease ergotism, with nausea and vomiting, and also to more serious neurological and vascular problems.

Antivitamins

Certain substances in food can act as antivitamins, inactivating vitamins or limiting their absorption in the human gut. The best described is thiaminase, present in certain fish. It has been shown that animals fed raw fish containing thiaminase can become thiamine deficient. It has not been clearly demonstrated that antivitamins are a major problem in humans. Haemorrhages have been observed in cattle that consumed feed containing dicoumarol, a substance that can have a negative impact on vitamin K and lead to bleeding.

Cassava Toxicity

Cassava is not indigenous to Africa, but it is widely used as a food in both East and West Africa, as well as in Asia and Latin America. It is usually eaten without any toxic effects, either because of the varieties used or because of local preparation measures that remove the toxin. Some types of cassava contain a cyanogenic glucoside which can result in acute toxicity with serious symptoms and death. It may cause nerve damage leading to paralysis, or it may behave as a goitrogen, aggravating iodine deficiency disorders (IDD) and causing goitres. In many African societies people know how to remove the toxin, mainly by soaking and sometimes also by grating and drying the cassava. Peeling cassava also helps remove the toxin. Toxicity occurs less frequently in Asia and the Americas.

Goitrogens

Some foods other than cassava contain substances that have been termed goitrogens, which appear to make those consuming them more likely to get goitre or IDD. The main goitrogens are thiocyanide, which reduces the levels of iodine in the thyroid gland, and thiouracil, which reduces the secretion of thyroid hormone. These goitrogens are most common in vegetables of the genus Brassica such as cabbage, cauliflower, mustard and rape.

Allergens in Food

Many people are allergic to one or more foods. Allergens vary in composition and in the foods in which they are found. Shellfish and other seafoods are especially common causes of allergic reactions.

Metals in Food

Industrialization, urbanization and the improper disposal of waste from factories and other businesses have caused metals which may be toxic to enter the food supply. A classic example is mercury in fish. In the early 1970s in the United States, various kinds of fish, such as swordfish, could not be sold because they had more than the permissible level of mercury, 0.5 parts per million (ppm). Mercury poisoning has also been a problem in fish in Japan.

Of much greater prevalence worldwide, especially in poor urban areas, is the problem of lead poisoning. Some of the lead consumed comes from foods, particularly animal foods such as meat and milk from animals that have consumed lead. Lead is also inhaled, for example from lead-containing fuels, and it can be ingested from water which has flowed through lead pipes and from the lead-based paints used in old houses. Lead poisoning causes long-term neurological problems, reduced psychological development in children and bone changes.

Agricultural Chemicals

The green revolution, which has led to higher yields of cereals and other agricultural advances, has enhanced farmers' ability to produce food in adequate amounts to feed the rapidly increasing population in the world. Some of the advances are dependent on the use of chemical pesticides, which are used to control weeds and a variety of pests, from marauding animals such as rodents, monkeys and elephants to disease-causing organisms such as parasites, moulds, fungi, bacteria and viruses. Farmers also use externally applied medications such as insecticides and oral or injectable medicines such as anthelmintics to rid their domestic animals of, for example, ticks on the skin and worms in the intestinal tract. These chemicals, their residues or metabolites may end up in the food that humans consume; some of them present health hazards. Textbooks of toxicology cover these in detail, and only a few are mentioned here.

The Joint FAO/WHO Expert Committee on Food Additives (JECFA) is responsible for reviewing the safety of residues of veterinary drugs in foods for human consumption and from time to time recommends safe limits. The Codex Alimentarius Commission can then adopt these limits as recommended international standards.

Under optimal agricultural and animal husbandry practices the residues of chemicals used would not present a risk either to agricultural workers or to consumers. Most countries have regulations regarding the permissible use of these chemicals. Some have monitoring systems. The efforts of the Joint FAO/WHO Meeting on Pesticide Residues (JMPR) have resulted in authoritative reviews of the safety of agricultural pesticides. JMPR has assessed the potential health problems from these chemicals based on the current literature and has recommended maximum residue limits both for adoption by the Codex Alimentarius Commission and for broad dissemination to member countries. In poor countries the regulations are often not adhered to and monitoring fails to detect many potential or actual problems.

In the use of farm pesticides, the first risk is to the agricultural workers who use them. They need to have dear instructions for the use of the chemicals. They need to know how to protect themselves, and they must have protective clothing and facilities to clean their bodies and clothes after working with pesticides.

Pesticides may also contaminate food. They are used in food storage to prevent spoilage or loss of food, and in this way too may present a danger to the consumer.

Most countries have regulations for pesticide residues in foods, and these need to be monitored and enforced. For example, the United States Environmental Protection Agency lists maximal residue levels of some 90 pesticides in foods sold for human consumption. DDT (dichloro-diphenyl-trichloro-ethane), which was used both for agriculture and also to kill mosquitoes in anti-malaria programmes, has been banned by many countries (and by all countries for use in agriculture), but others have felt that the risk from malaria was greater than the risk of toxicity from DDT. Now there is greater concern for other insecticides. Of particular concern now are polychlorinated biphenyls (PCBs); the organophosphorus pesticides such as malathion and parathion, widely used in agriculture; dieldrin; and the herbicide chlorophenoxy acid. In most countries the Acceptable Daily Intake (ADI), set by FAO and WHO (via JMPR), is the standard for monitoring.

Although there have been a few industrial accidents involving workers accidentally being sprayed with pesticides and an occasional poisoning of a child who drank a pesticide solution, both are rare. Extremely few cases of pesticide poisoning from eating food have come to the attention of JMPR.

Food Additives

Chemical or other substances are added to foods for human consumption for many different reasons. The most important perhaps is to preserve the food, but additives may also be used to change the colour, the taste or some other quality of the food. Some countries have very strict regulations governing the approval of a new food additive for use by the food industry. For those additives that are approved, the regulations usually state the maximum level that is permitted. Again it is JECFA that has established the safe levels which then have been used by the Codex Alimentarius Commission. Concerns regarding food additives are that they might be carcinogenic (stimulate cancer) or have a negative impact, including genetic or teratogenic effects, on the foetus if consumed by pregnant women. In the United States food additives approved for use by industry are listed as "Generally recognized as safe" or GRAS. The GRAS list includes many additives in use before 1958 which evidence suggested were safe, and new items introduced since 1958 which are rigorously tested to show among other things that even rather large amounts are not carcinogenic in laboratory animals. JECFA has prepared specifications for food additives which guide Member Governments to establish the identity and quality of additives being used. These specifications are also used by industry.

Radioactive Contamination of Foods

Happily, the contamination of foods with radioactive fallout, either from explosions of atomic bombs or from accidents at nuclear power plants, is rare. The accident at Chernobyl in the former Soviet Union in 1986 was the worst well-described accident of this kind. When radioactive dust is liberated into the atmosphere, it is blown by the wind and falls to earth where it may contaminate food crops such as cereals, fruits and vegetables, but also grass which is then eaten by cattle and other livestock. As a result the milk and the meat of these domestic animals may contain unacceptable levels of radioactive materials. Following the Chernobyl accident, elevated levels of diseases such as thyroid cancer (presumably because of fallout of radioactive iodine, ^{131}I), particularly in children, and other malignancies have been reported.

Very soon after the Chernobyl accident FAO convened an expert consultation which recommended action levels for radionuclide contamination of food in international trade. No guidelines existed previously, so this rapid action was important. In the event of a nuclear disaster, people living in the area of the fallout should avoid eating foods that were growing in the affected area. They should also avoid consuming milk and meat produced in the area and foods that might have been exposed to dust fallout. Foods stored in sealed containers, including tins, are safe. The authorities should bring food into the area from unaffected areas as soon as possible. People all over the world should be made aware that a nuclear accident has occurred and that it could make their usual food dangerous.

Consumer Protection

Many of the actions discussed earlier in this chapter will help protect the consumer and ensure a safe diet of good quality. Some other specific activities might further help the consumer. In many countries greater attention is now being given to food labelling, which may be controlled by regulations. FAO has had a leading role.

Recommended nutrient reference values for labelling purposes were established at an FAO consultation in 1988. Food packaging that provides useful information for the consumer can be helpful. It should, if possible, express in simple terms the amount of nutrients in the food, perhaps as percentages of the requirements or allowances of each important nutrient per serving of the food. The energy, protein, carbohydrate and fat content of one serving should also be included. In countries where there is concern about arteriosclerotic heart disease, this information might be broken down further to indicate the amounts of different kinds of fat, cholesterol and fibre. Food labelling might also include the amount of additives in the food.

Food advertising should use only truthful information and should not make claims for the food that are not true. Foods that can be harmful should perhaps not be advertised.

In 1981 at the World Health Assembly in Geneva, 118 nations voted in favour of adoption of the International Code of Marketing of Breast-milk Substitutes, which calls for the cessation of all promotional advertising of breastmilk substitutes to the public. (Only one country, the United States, voted against the code.) Many countries have introduced legislation to limit the promotion of infant formula, because it is generally agreed that breastfeeding is very important for good health and good nutrition and that promotion of infant formula has greatly eroded and undermined breastfeeding. The multinational corporations that manufacture infant formula continue strenuously to promote infant formula in other ways than advertising to the public, for example by offering free samples and by providing literature to the medical profession.

Improving Food Quality and Safety

Most countries have legislation to help ensure the safety, and sometimes the quality, of foods from production to retail sale. Farmers, food processors and the public, however, are not always conversant with the regulations. Moreover, dishonest traders may seek to ignore the regulations. As a result, unsafe foods that are contaminated or spoiled or have dangerous levels of chemicals are reaching consumers, sometimes widely, putting the public at risk of illness.

Most countries have established institutions or branches of ministries (such as a bureau of standards or a branch of the ministry of agriculture) designated to ensure food quality and safety. These mechanisms often need strengthening, and there are often too few well-trained people or well-equipped laboratories to monitor the situation. In some countries an interdisciplinary, interministerial committee could be established to examine all areas related to food safety, to ensure that the most important aspects get covered and perhaps to suggest the most important and most feasible priority areas. Such a committee could have many functions, but the main ones might be to promulgate and implement food safety standards; to establish means of monitoring, including inspection, sampling and testing; to recommend a programme for educating both food industry personnel and the public on food safety; and to find ways to involve and obtain assistance from international agencies such as FAO and WHO (with the Codex Alimentarius Commission) and other foreign institutions.

The following actions might be given priority for immediate strengthening because they would cost little and seem feasible and important:

— greater attention to food hygiene practices at all places where cooked food is sold to the public (restaurants, stalls, street carts);

— education of employers and employees in the food processing industry and monitoring of their food safety practices nationwide;

— ensuring improved meat inspection at all abattoirs;

— collaboration with the ministry of education to produce a syllabus and perhaps a booklet on food hygiene for primary and secondary schools;

— seeking fellowships to send university graduates overseas for high-level training in food science with emphasis on food safety;

— teaching specific steps to improve the quality and especially the hygiene of street foods.

Steps to improve the quality and safety of food are important if people are to have good health and nutrition in developing countries. Such steps will also benefit food trade because contaminated or unsafe food should not be traded either in internal markets or for export. The FAO/WHO Codex Alimentarius Commission can assist non-industrialized countries in implementing standards and codes with the objective of protecting consumers and promoting food trade. FAO can assist governments in modernizing their food regulations, in designing compliance systems, in training food inspectors and related personnel, in improving food analysis laboratories and training their staff and in actions to ensure better quality control by food producers, manufacturers and processors. Food quality needs to be protected from the farm to the consumer. FAO, with WHO, can also help with the scientific evaluation of food additives, various contaminants and medicinal products. In the years ahead countries will need to give consideration to the agreements of the General Agreement on Tariffs and Trade (GATT) regarding sanitary, technical and other regulations which may be barriers to food trade.

In conclusion, consumers have a right to expect that their food is safe and of good quality, and both the food industry and governments have a responsibility to honour that right. To do so will require knowledge on food safety on the part of farmers, food processors and the public plus effective food safety control activities by the food industry and government. Control of food safety requires that there be in place laws, regulations and standards related to food quality and safety plus a system for food inspection and for monitoring to ensure compliance. Some inspection and monitoring can be achieved without extensive facilities, but there will be a need for laboratories to undertake the important analyses recommended. International agencies such as FAO may be called upon for technical and other assistance. FAO's very important efforts in helping to establish and strengthen food control systems internationally and particularly in member countries must be recognized. The Organization's work and actions over many years have contributed substantially to significant improvements in the overall quality and safety of the food consumed in many countries, especially in developing countries worldwide.

References

FAO. 1992. *Integrating diet quality and food safety into food security programmes,* by M.F. Zeitlin & L.V. Brown. Nutrition Consultants' Reports Series No. 91. Rome.

FAO/WHO/United Nations Environment Programme (UNEP). 1990. *Manuals of food quality control. Food inspection.* Rome.

McLaren, D.S., Burmad, D., Belton, N.R. & Williams, N.F. 1991. *Textbook of paediatric nutrition.* Edinburgh, Scotland, UK, Churchill Livingstone. 3rd ed.

Passmore, R. & Eastwood, M.A. 1986. *Davidson and Passmore human nutrition and dietetics.* Edinburgh, Scotland, UK, Churchill Livingstone. 8th ed.

7

Household Food Security

Food security is frequently defined as access by all people at all times to the food they need for an active and healthy life. Household food security, in turn, means adequate access by the household to amounts of food of the right quality to satisfy the dietary needs of all of its members throughout the year. A family can secure food in two main ways: food production and food purchase. Both require adequate resources or income. Other less important, less common ways of obtaining food are through food gifts or charitable or government food allocations, in free school meals or with food stamps. The importance of agricultural food production to underpin national and local food security was outlined. Food security was shown to be important at all levels, but particularly at the household level.

Food security for the individual child (or for the family) is one of the three essential ingredients in preventing malnutrition. Individual food security is essential for good nutrition, but it does not ensure good nutritional status, because other factors such as disease, infrequent feeding, lack of care and poor appetite may adversely influence nutrition.

The achievement of food security requires:

- a sufficient supply of food;
- stability in the supply of food, both throughout the year and from year to year;
- access, both physical and economic, to food, which requires the ability and wherewithal to produce or procure all the food needed by the household and each of its members.

The main underlying determinant of household food insecurity is poverty. In Asia, Africa and Latin America a large proportion of the population in both urban and rural areas is affected. It has been stated that not all poor people are undernourished, but most undernourished people are poor.

Household food security in each country, even if the country is food secure, depends partly on the extent to which the country pushes for greater equity in incomes, in land distribution and in access to services. National policies may not only help farmers achieve increased food production, but may also help people satisfy their food demands. Although household food security is most influenced by actions at the household level, factors and actions at the local, national and international levels also have effects.

Forms of Food Insecurity

Household food insecurity takes different forms which require different responses or actions. The approaches may be different depending on whether food insecurity is chronic (with households almost always short of food) or transitory (resulting from temporary adverse circumstances). Food insecurity may be seasonal; a family may have insufficient food perhaps each year or most years, but only in certain seasons.

The consequences of household food insecurity are as different as the causes. Which members of the household are most affected will vary, sometimes as a result of intrahousehold food distribution. Thus two families each with a mother, a father and two young children, with similar moderate but not severe insecurity, may respond in different ways, with different outcomes. The first family may believe in "children first" and despite food shortages may make certain that the two children receive all the food necessary for good growth and health; the adults then may develop signs of undernutrition or more likely will reduce their energy expenditure by reducing their activities and productivity. In the second family, the father may always satisfy his desires for food first, leaving the remaining food for the mother and, last, the two children, who get less food than required. In this family the children would show evidence of malnutrition. Sometimes, however, ensuring the energy and nutritive intake of the food producer and wage earner may be necessary for the family to get the food it needs for survival.

Households most likely to be food insecure, or at high risk of food insecurity, are the poorest. In rural areas these may be landless households; those with such small plots of land (sometimes marginal land) in relation to family size as to make adequate agricultural production impossible; sharecroppers or tenant farmers who get relatively little of the crop produced; pastoralists, fishermen, forestry workers and others who earn too little money or produce too little food for the needs of their families; female-headed households where the mother has many responsibilities for child care as well as farming; and poor households with a high dependency ratio or that have no or few active adults because of age, disease, disabilities or other reasons.

In urban areas also the most food insecure are the very poor, including households where there is unemployment or underemployment; single female-headed households with dependent children; elderly people living alone; destitute and homeless individuals; and those with chronic debilitating disease or serious disabilities.

Increasingly the acquired immunodeficiency syndrome (AIDS) epidemic is contributing to food insecurity, sometimes because adults who were breadwinners have become seriously ill or because orphaned children as young as 12 years of age have become household heads caring for younger children. In addition, where human immunodeficiency virus (HIV) infection is prevalent, the disease is having a major negative impact on agricultural production, on economics and on health services.

Issues in Household Food Security

Many variables influence household food security, and all of them can be manipulated to some degree to improve it. However, there are few easy answers or prescriptions for alleviating food insecurity. Recommendations often depend on local circumstances. Solutions will almost always involve participation at the local and household level.

Among the issues that influence household food security are adequacy of local food supplies; potential for cash crops and home gardens; urban versus rural food supplies; producer and consumer prices; available means of improving food production; food storage and stabilization of food supplies; employment questions; and labour-intensive versus labour-saving work. Agricultural and planning ministries and other organizations need to address some of these issues at the national level.

Other issues of great importance to food security involve gender. What roles do males and females have in the society? To what extent are females discriminated against? Do women have an unfair labour burden? Who controls household finances?

People have different ways of dealing with food insecurity depending on their systems of earning a livelihood or procuring needed food. There are major differences between subsistence farmers and pastoralists; between sharecroppers and urban workers; and between welfare recipients and those working in the informal economy. Clearly urbanization and migration from rural areas have a role in food security.

Evidence about household food insecurity and its causes strongly suggests that in many instances attempts to improve food security should start not at the national level alone -the classical approach - but at the household level, or preferably at both. Emphasis must be on local-level planning of community interventions and on using a participatory approach.

The three important requisites of household food security - an adequate local supply of food, stability in the food available and accessibility of food - are discussed below. For nutritional security there must also be adequate health and adequate care, and the food must provide all the nutrients needed for good nutrition.

Food Supply

If there is an insufficient quantity of food to meet the food needs of a population, then some persons or some households will be food insecure. Various steps along the food chain need to be considered in relation to supply.

To improve household food security, various methods of increasing sustainable agricultural production of food (or other methods of food acquisition) need to be promoted. It is also necessary to ensure good harvesting and storage of food with the smallest possible losses; an effective and efficient marketing system; and good food processing and preparation. All these topics are discussed in detail in many publications, of which some are listed in the Bibliography.

At the national level food supply also depends in part on government and private-sector decisions and actions concerning what amd how much food to import and export, when to do so and how to allocate resources.

These decisions in turn depend on whether domestic food production is able to meet local needs. If imports are necessary, the amounts and types of food imported will depend on many factors including political considerations, availability of funds and foreign exchange, trade policies, world food prices and perhaps availability of food aid.

Often government economists and planners, considering the supply side of food security, address only the need for adequate energy for the population in terms of cereal grains and perhaps legumes. For good nutritional status, however, the production, supply and availability of other foods, including fruits and vegetables, need consideration.

Stability of Food Supplies

A reasonable degree of stability in the supply of food during the year and in all years is a necessary ingredient of food security. This stability can be ensured in various ways, including:

— adequate stockholding by ensuring strategic food reserves;

— a good food marketing system at all levels, including the village level, throughout the year;

— protecting or introducing a variety of cropping strategies such as mixed cropping, proper rotation and use of appropriate agricultural inputs;

— promoting good post-harvest food handling, transportation, distribution, preservation, storage and safety;

— assisting where appropriate with increased production of fish and animal products for human consumption (which includes attention to animal health);

— promoting household, school and community gardens, especially stressing production of fruits and vegetables;

— ensuring the sustainability of food supplies using agricultural, industrial and marketing strategies and relying on renewable resources with proper concern for the environment.

Access to Food

Household food security depends on access by all household members to food that satisfies their nutritional needs at all times. Each household needs to have the resources, the ability and the knowledge to produce or to procure the foods that it needs to provide for the energy needs and the nutrient requirements of every member. It is important that households be able to acquire adequate quantities of food all year and in all years. The food must be culturally acceptable.

The acquisition of adequate food depends on how much a person, family or household:

— owns (land, resources, etc.);

— produces;

— receives (gifts, government aid, charities, etc.);

— trades, barters or exchanges;

— inherits.

There are obvious differences in how urban dwellers and rural farmers usually obtain access to sufficient food for themselves and their families. Most urban households usually need to obtain sufficient money to purchase enough food to satisfy the nutritional requirements of all their household members. By contrast, the rural landowner or farmer must have enough land, resources and labour to produce sufficient food to feed all household members or to sell for cash with which to buy the ingredients of an adequate diet for all. The rural family that has neither land nor labour usually needs to obtain enough money to purchase food, much as urban households do. Many farming households are also dependent on off-farm income-earning opportunities.

Where household insecurity is prevalent among both urban and rural people, attention has to be given to ensuring that farmers are paid remunerative prices for their produce; that processing and distribution systems are expanded and efficient; that minimum wages are adequate; that prices of staple and perhaps other important foods are reasonable, or even subsidized; and that other essentials (such as housing, health care, education and transport) are affordable for those receiving the minimum wage. Programmes that provide social security, welfare and unemployment payments or that provide free or subsidized food (through food stamps or school feeding, for example) will help the poor and disadvantaged obtain access to food.

Rural farming households can take measures, and authorities can help them, to optimize production from their land and to get the most food and money from farm production. In some parts of the world implementing land reform policies to allocate adequate land to poor rural families and ending sharecropping might help families become food secure. In many areas livestock are integral components of farming systems and may provide insurance against bad agricultural years, as a form of asset that can be exchanged for money to purchase food. Rural families may also be assisted with credit, subsidized foods, food stamps or charitable help, especially in bad agricultural years.

It has been observed that where there is food shortage and famine, families with money and resources do not suffer from starvation. Very poor families have the fewest assets and thus are usually the most food insecure and the most vulnerable to serious food crisis.

Responsibilities for the Right to Food

Sustained improvements of household food security are likely to depend on actions at the local and household level and on the participation of the poor in bettering their own lives. However, this idea should not allow those who are better off to forget that adequate nutrition is a basic human right and that the occurrence of malnutrition among so many people in the world is an indictment of all who permit it. The world is divided into nation States, each of which has a major influence on its own inhabitants. Each State has the responsibility to respect, protect and fulfil human rights, including the right to adequate food and nutrition. Respecting means that the State does not take actions or have policies that make it more difficult to procure food for the needs of its people, even in times of crisis or conflict. Two examples of protection would be preventing individuals from being deprived of their abilities to produce food or to earn money to purchase food; and establishing and enforcing regulations to ensure consumers a safe food supply. Fulfilment of the right to food and adequate nutrition includes the State's obligation to provide assistance to the vulnerable to meet their food needs even at times of shock or crisis.

Meeting Desirable Allowances for Energy

It cannot be stressed enough that humans have a right to sufficient food and good nutritional status. "Sufficient" food, moreover, must provide not only for basic energy requirements, but also for the energy requirements of an active and healthy life.

The concept of desirable allowances has important implications for improving food security. If only energy requirements or needs are considered, a person in energy balance who is not clinically undernourished or does not have low body mass index (BMI) or low weight for height might be considered food secure. However, such a person may be foregoing desired activities to conserve energy. That person has unfulfilled energy wants, does not have enough energy to satisfy desirable allowances and is food insecure.

Energy balance is not an indicator of adequate energy intake. A person may be in energy balance, with energy intake equalling energy expenditure, but may in fact be greatly reducing his or her activity levels to remain in balance. Consciously or unconsciously he or she may choose to do less work on the farm, to reduce household chores, to play less with the children, to refrain from participating in sports and to curtail social and community activities, and instead may rest more and sleep more. This individual, though in energy balance, is nonetheless in a state of energy deprivation and is therefore not food secure; yet a physical examination may indicate no physical evidence of malnutrition or undernutrition.

There is a difference between affluent and poor people who are in energy balance over time. Generally, affluent people adjust energy and food intake to meet energy expenditures, while very poor people with food shortages adjust their activity to their energy intake (foregoing activities to conserve energy). Where food is plentiful, people may forego food or increase exercise to maintain balance (the jogger's syndrome); where food supply is deficient, they may forego activities to maintain balance.

Very little research has been conducted to determine what activities are foregone, and to what extent, in order to maintain energy balance when too little food is available. Policies and programmes are needed to ensure that energy wants as well as energy requirements are satisfied. Providing women with the opportunity and freedom to control their own fertility is also important. These issues are all highly relevant to any consideration of food security.

Indicators of Household Food Security

As stated, adequate supply, stable availability and proper access to food are essential requirements of household food security. Indicators of household food security are then those related to food production and supply on the one hand, and food demand and access on the other. Manuals and books have been written on agricultural production, nutrition surveys, food balance sheets, household economics and other topics related to specific indicators of food security. Here a few key indicators are briefly mentioned.

Indicators related to food supply include:

- measurements of agricultural production (similar to those collected for food balance sheets);
- inputs that influence agricultural production in the area or country (such as credit, irrigation, fertilizers and pesticides);
- climatic data (especially the amount of rainfall compared with that usually expected and the timing of the rainfall, but also temperature and other meteorological data);
- market factors, including food sales and prices;
- security (whether there are areas of conflict or parts of the country where movement of people and food is restricted or limited);
- data on crop diseases and agricultural pests.

Indicators related mainly to household access to food include:

- food consumption data;
- clinical assessment related to signs of nutrient deficiencies;
- anthropometric data such as BMI;
- assessment of food stores;
- selling of assets (or loans obtained for assets) including livestock and household goods;
- greater consumption of low-status foods (a move from rice to cassava consumption, for example);
- increases in food foraging and gathering of wild foods;
- migration from rural to urban areas;
- data suggesting frequent perception of food insecurity or food crisis by household members.

In many countries with diverse topography, agricultural conditions and peoples, indicators may need to be rather specific for particular areas of a country or particular groups of the population.

Nutritional surveillance may be established as a system for regular monitoring of the food situation, the functioning of the food system and also some aspects of the nutritional status of the population. This system will then, depending on the data being collected, provide indicators of household food security. Sometimes nutritional surveillance is established as an early warning system to help predict serious food shortages and to trigger action. Some countries have established nutritional surveillance as a means of providing data to influence government policies.

Coping at the Household Level

Often poor households have amazing resilience and an impressive ability to cope with short-term crises and to survive on low incomes and what appear to be relatively low availabilities of food.

Transitory or short-term food insecurity is often the result of a shock that has struck a blow to the household. The coping mechanisms adopted depend partly on the nature of the shock and partly on the household's circumstances. Different members of the household may respond to a shock in different ways. It has been suggested that there are four main types of shocks:

- work shocks, when there is a sudden fall in the availability or the amount of work on which a family is dependent for income, or a fall in the wage rate;
- output shocks, when the production output of different family members or the money received for a given output from work declines suddenly and significantly;
- food shocks, when food in the marketplace is less available and/or when food prices rise, either of which results in less food in the household;
- asset shocks, when the assets of the household are reduced in quantity or value because of fire, theft, the death of livestock or small animals owned by the family, inflation or sale of assets to raise money.

Where shocks cause transitory food insecurity, and also when families face chronic food insecurity, families take actions to ensure adequate food. Examples of such actions include:

- using labour differently, perhaps having one or more family members move from the rural area to town to earn money;
- using money differently, perhaps by purchasing cheaper foods (such as yams in place of bread or cassava in place of rice) and foregoing non-food purchases (e.g. not buying school uniforms or not paying school fees);

- purchasing less kerosene (for lighting or other uses);
- selling or pawning household assets (farm animals, bicycles or luxury items such as watches and radios);
- securing credit or loans, which for the poor is often very difficult;
- entering the informal economy, either legally (having older children shine shoes or clean automobiles) or illegally (through prostitution or thieving);
- seeking assistance from government or non-government programmes (e.g. feeding programmes, food subsidies or food-for-work programmes).

Imaginative programmes to assist poor families to overcome the results of shocks will help reduce food insecurity.

Government Actions to Improve Household Food Security

Food security may be influenced by anything that governments do to improve income and reduce poverty; to increase agricultural production, especially by poor rural families; to ensure prices that are fair to producers and consumers; and to make services available to people.

Some examples of more specific government actions include:

- increasing agricultural food production, preferably using sustainable methods, in such a way that poor subsistence farmers who are most vulnerable to food insecurity derive benefit;
- taking steps, if necessary, to import more food and to limit the export of food where this will improve food security;
- promoting improved marketing of food and better food distribution in a manner that addresses problems of food insecurity;
- in times of modest food crises when shortages of food are predicted, releasing or moving foods into the crisis area to prevent price rises and profiteering and to stabilize supplies so that market mechanisms protect the poorer people from a food shock;
- if the preceding strategy is not working or is deemed impossible, effecting food price controls, subsidization or rationing, but only if doing so is unlikely to be a disincentive to food production;
- improving equity by ensuring that all people pay a fair share of taxes and perhaps by increasing minimum wages and offering subsidized or free services to poorer people;

— streamlining the purchase of cash crops produced by small farmers so that most of what is paid goes into their pockets rather than lining the pockets of intermediaries or being spent on bureaucratic marketing practices.

Besides these specific measures, governments need to have a sound overall development strategy which creates conditions for economic growth with equity. Poverty alleviation programmes need to be sustainable. The rate of exchange of currency in a country, the nation's export and import policies, the rate of inflation, the budget deficit and debt repayment obligations can all influence prices, unemployment rates and incomes of the poor. Much recent discussion has focused on structural adjustment programmes, sometimes mandated for poor countries to promote economic growth. These programmes have caused major problems for the poor, often through reductions in producer and sometimes consumer subsidies. Of great concern also has been the reduction in social services: previously, many countries in the South had free primary and secondary education and free health services, including both out-patient care and hospitalization, but by 1992, with implementation of structural adjustment and sometimes for other reasons, school fees and charges for health services had become common or even the norm. These changes have had a marked impact on the poor, in some cases worsening the problem of food insecurity.

In some countries, particularly in Asia and Latin America, economic development has progressed and the creation of wealth has resulted in a reduction of malnutrition and a lowering of infant mortality rates. However, in other countries, particularly in Africa, economic policies, in concert with very adverse socio-economic and ecological conditions, appear to have sometimes aggravated malnutrition. When this outcome can be predicted, governments need to consider taking early measures to compensate for likely adverse effects, to lessen the hardships for the poor.

Promoting rural development with a special focus on sustained reduction of poverty among the rural poor can improve food security. Appropriate technologies and sometimes producer incentives to increase both production and employment in rural areas can help to reduce food insecurity and poverty. These strategies need to be imaginative and innovative, but there have been successes which give grounds for optimism. For example, credit is often a serious problem for the rural poor, but the Grameen Bank in Bangladesh has made thousands of loans to poor people, many of them female-headed households. The bank has achieved a good record of repayment and has helped lift many people out of poverty. Agricultural extension aimed at poorer farmers has moved research results from universities and research institutions into the fields of poor farmers. Attitudes in many countries have changed, so that strengthening local leadership and empowering women are on the agenda of many countries. Participation and community involvement are increasing rapidly. Non-governmental organizations (NGOs) which work well with people at the local level are absorbing external funds that previously were poorly utilized by major government or international agencies.

Agrarian reform remains a problem, particularly in certain Latin American and Asian countries. In several countries, lack of tenancy reform, continuing bias against females and social discrimination, including caste differentials, still contribute very significantly to food insecurity. Redistribution of land is still much needed. Lower-caste families need to have full access to all services. In certain countries, such as Indonesia, the strategy of resettlement on new lands, often on less-populated islands, can reduce food insecurity.

Government and the private sector can reduce poverty by increasing employment opportunities in both rural and urban areas. They should aim to improve both the incomes of the poor and also where possible their capacity to earn income. Some governments can invest in public works, particularly labour-intensive ones, and in programmes focused on parts of the country that have high rates of poverty.

References

Brun, T.A. & Latham, M.C. 1990. *Maldevelopment and malnutrition.* World Food Issues, Vol. 2. Ithaca, New York, USA, Cornell University, Program in International Agriculture.

Dawson, R.J. & Canet, C. 1991. International activities in street foods. *Food Control,* 2(3): 135-139.

Maxwell, S. & Frankenberger, T.R. 1992. *Household food security: concepts, indicators, measurements. A technical review.* New York, USA, United Nations Children's Fund (UNICEF)/International Fund for Agricultural Development (IFAD).

FAO. 1990. *Women in agricultural development. Gender issues in rural food security in developing countries.* Rome.

8

Healthy Diets and Lifestyles

Very young children depend for their nutrition on good care. Of course, everyone benefits from care: health, nutrition, and general well-being blossom in a caring environment. Clearly very young children, certain older people, some sick people and some physically or mentally ill people are especially dependent on care. For young children the relationship between care and nutrition is especially strong.

Infants and young children up to age three years are almost totally dependent on others for food and therefore for good nutrition. Children three to five years of age have some ability to gather food, to select a diet and to feed themselves, but in most societies children up to the age of about six years, or school age, would also be considered in need of feeding care. Thereafter, care is highly desirable but not essential for survival. However, good care will always positively influence nutritional status and well-being.

Of the three underlying causes of malnutrition, namely food, health and care, the one with the least investigated and the least understood role is care. Food security and health have long been known to have an important relation to nutrition, and a huge literature and extensive range of interventions focus on them. Few programmes designed to improve nutrition include a set of actions to address problems related to care.

The English word "care" is both a verb and a noun. In The Oxford English dictionary definitions of the verb include to feel concern or interest, to provide food, attendance, etc. for (children, invalids, etc.), to look after and to provide for, and meanings for the noun include solicitude, anxiety, serious attention, heed, caution, charge and protection. Engle provided a working definition referring to the care of young children: "Care refers to caregiving behaviours such as breastfeeding, diagnosing illnesses, determining when a child is ready for supplementary feeding, stimulating language and other cognitive capacities and providing emotional support".

In most developing countries the mother is usually the main care-giver for the infant and the very young child, but in the common extended family grandmothers, siblings, the father, other family members and people outside the family often contribute to child care. As the child gets older care may be increasingly provided outside the home, for example, in day care facilities.

Adequate care is important not only for the child's survival but also for optimal physical and mental development and good health. Care also contributes to the child's general well-being and happiness, otherwise termed a good quality of life. Care influences the child and the child influences the care.

The inadequate food, health and care which lead to malnutrition can be factors at the international, national, local and family level. Child care may be influenced by international factors such as war, blockade or global determinants that keep nations in poverty; national factors such as equity issues and availability of good health services and education; local factors such as land distribution, climate, water supplies and primary health care; and family factors such as the presence of other family members, type of housing, availability of water, household hygiene and mother's knowledge.

Promotion of Good Caring Practices

Care-giving behaviours that contribute to the good nutrition, health and well-being of the child vary enormously from society to society and from culture to culture. A first assumption can be made that almost all societies value children and wish to see them grow to be healthy, intelligent and productive adults. A second assumption, which is more debatable, is that societies in general have traditional or culturally determined caring practices of which most are good and contribute well to child development, including good nutritional status.

Besides these two assumptions, it is submitted that in Africa as well as in most of Asia and Latin America, problems with good child care in the 1990s may be related more to an erosion of traditional caring practices than to the fact that important caring practices in the society were, or are, wrong or inappropriate, or important contributors to malnutrition. (There are exceptions; for example, a traditional caring practice that has been an important contributor to malnutrition is the favouring, in terms of diet, health and care, of male over female children in some areas of South Asia.) Traditional caring practices in their broadest terms have been altered, often for the worse, as a result of modernization, westernization and increasing urbanization. A good example, and the one most written about, is the decline of breastfeeding, which was a good traditional practice almost everywhere. Its decline has in large part been influenced by modernization, including promotion by infant formula manufacturers and the medical practices of Western-oriented health professionals.

Protection of Good Practices

Protecting is an essential part of any strategy for optimal care to ensure good nutritional status. Good practices need to be protected from erosion by many different factors. For example, in a society where most mothers breastfeed their babies for 18 months or longer, with no or few other foods introduced until the child is four to six months of age, protection should take priority over support and promotion of breastfeeding. Similarly, protection is warranted if a society traditionally provides a lot of stimulation for children; if the infant is seldom left alone but is carried on the mother's back; if fathers, grandmothers, older siblings and other relatives frequently help in child care; and if traditional weaning foods of groundnuts, green leafy vegetables and legumes with a local cereal gruel are the norm. These practices may be threatened by modern or Western influences. A new television set in the family may result in adults neglecting to stimulate their children; advertising and promotion of expensive manufactured weaning foods may lead families to poorer diets at higher cost; or work away from home may cause long separations of the mother and her infant.

Support

Support is particularly appropriate when mothers' or families' good traditional caring practices are threatened or being eroded because of changes in society, which may result from modernization, westernization or urbanization. Support includes activities, both formal and informal, that may help women in changing circumstances to follow those good caring practices that were once considered normal and are now threatened. Support may involve restoring confidence in mothers, strengthening their belief that traditional good caring practices may be better than new practices that may seem modern and up to date but are in fact inferior. For example, westernization and modernization may suggest that a modern woman does not breastfeed her baby in a public place; that canned baby foods are superior to home-prepared foods; that salt and sugar comprise a better treatment for mild diarrhoea than family soups and breastfeeding; that it is better for a child to stay at home and watch television than to go with the mother to the village market; and that eating with a fork is preferable to eating by hand after traditional hand-washing. In fact none of these "modern" practices is better for the child than the traditional alternatives.

In many developing countries paid employment for women away from home is an important factor in the erosion of traditional good caring practices. It has certainly made breastfeeding more difficult. Three months of maternity leave would help support mothers in providing initial infant care. Then, during the eight hours mothers are away from home, a crèche or day care centre at the place of work would be supportive. Support for good traditional care may include mothers' support groups or arrangements for adequate child care while the mother is away from home. Staggered working hours for different family members and a greater role for the father in child care could also help.

Promotion

Promotion is particularly important when some, many or most good traditional caring practices have been abandoned or lost. Promotion involves motivation or reeducation of mothers, other family members or whole communities. It is the most difficult and the most expensive of the three strategies.

It may be important to start by identifying the most important factors that led to the decline or disappearance of good caring practices. There must be evidence that the new caring practice is less desirable and less beneficial. A lack of such knowledge will almost certainly lead to failure of a promotional campaign. Properly applied social marketing methods and techniques may be useful. Political commitment and will may be necessary. The promotion of good caring practices will often involve public education and mass media efforts.

Some of the best examples of promotion of a good traditional caring practice that had been seriously eroded concern breastfeeding where it had markedly declined and been replaced with infant formula and bottle-feeding. Promotional campaigns in Brazil in the 1970s and in Honduras in the 1980s proved successful. Other practices for which promotion might be attempted include traditional breastfeeding and family feeding for children with diarrhoea; the carrying of children on the back of the mother where this has been replaced by leaving the child at home; and the use of good home-prepared weaning foods in place of expensive, less-nutritious manufactured foods.

Identifying Good Caring Practices

Mothers, fathers, families and communities (as well as governments and international institutions) take actions all the time that influence nutrition. These actions are in the area of food, health or care. They are based on, or arise from, everyday decisions. They may have a positive or negative influence, or they may be neutral.

The first step in making decisions that will lead to actions to protect, support and promote good child care is to assess current caring practices that may influence nutrition. For many countries where there is fairly good knowledge about the food situation and about health status and health care, there may be very few published findings on child care, especially as it relates to nutrition. There will often be some information on breastfeeding and weaning practices, but there are usually very few data or even descriptions concerning caring practices that influence psycho-social and motor development, maternal factors such as mothers' self-esteem and mothers' beliefs and attitudes regarding child care, or household and community factors that greatly influence child care. There may be ways of obtaining such information rather quickly; this may be the first activity, and it is an important one.

A useful approach for identifying child-caring practices that seem to be desirable may be an investigation of "positive deviants" in a community. Positive deviants are young children who have good nutritional status even though they come from very poor households, have uneducated mothers, have limited access to food and health services and live in a community where most children have malnutrition. If it can be found that the mothers and families of positive deviants have a set of caring practices not usually used by other families, then it can be assumed that all or some of these caring practices are good and deserve protection, support and promotion. A comparison of negative deviants and positive deviants may also be useful.

Actions in Favour of Good Care to Ensure Good Nutrition

Actions in favour of good care can be divided into three groups: delivery of services, capacity-building and empowerment. All three can operate at different levels in society (from national to family), and each contributes to the others.

Delivery of services in support of child care may address the most immediate causes and may sometimes be curative rather than preventive; examples include oral rehydration for diarrhoea, deworming and child feeding targeted at malnourished children. In other cases delivery of services may address problems from the top down and may be preventive to some degree; examples include immunization and organized day care centres. It should be accepted that delivery of services may not be sustainable or if sustainable may have to remain in place for a long period unless other changes prevent or permanently cure the problem in society, not just in the individual child. Oral rehydration prevents death in a child and treats dehydration, but it does not reduce prevalence or incidence of diarrhoea in society. Acknowledging the limitations of an action is just as important for its effectiveness as recognizing its successes.

The next level of action, capacity-building, is aimed to deal not with the immediate causes but more with the underlying causes of malnutrition. Consequently actions at this level are often preventive rather than curative and are likely to be more sustainable. These actions are also likely to be most successful if they work mainly from the bottom up, not from the top down. Capacity-building is seen as of very great importance for improved care in relation to nutrition and may involve protection, support and promotion. Examples include infant feeding practices that permit a smooth transition from exclusive breastfeeding to mixed feeding to exclusive feeding of home foods; child care practices that are stimulating and influence good psycho-social development; health education to provide knowledge about protection against disease; and home hygiene and sanitation to prevent diarrhoea and intestinal parasitic infections.

The third level, empowerment, crosses the boundaries of service delivery and especially of capacity-building. In general, however, actions that are empowering for mothers often address the more basic causes of child malnutrition. Empowerment for women involves ensuring that they have rights that women in many societies lack. Every woman everywhere should have the right to earn income; not to be overburdened with work; to breastfeed freely and easily; and to have reasonable access to services and resources and to capacity-building activities. Possible actions at the level of empowerment include those that improve mothers' income or control of family income; providing good access to health care for women and their children; managing water supplies to lessen the burden on women; and also many activities that reduce poverty and increase equity (including some trade and price policies). Some actions that are empowering are top-down and others are bottom-up activities.

Investigations on current good caring practices, on how they might be threatened by new influences and on how they might be protected in changing, modernizing, urbanizing societies deserve a very high priority. Support for good caring practices is undoubtedly also an important action, but it is perhaps not such a high priority for research, although some investigations will be needed.

Relatively little is known about which good caring practices that are not now the norm for particular families should be promoted or how to promote them. Where caring practices are inadequate and are causes of malnutrition, studies are needed on appropriate alternatives, how they might be promoted and their potential impact on child nutrition.

Some research has been published on intrafamily food distribution, meal frequency, energy density of foods and some other practical topics; but very little is known about some other important subjects that are related to care and that may influence nutrition.

Promoting Appropriate Diets and Healthy Lifestyles

The major nutritional problems in the world can be divided into two general categories:

— those caused by insufficient intake of nutrients, which may be related to food insecurity, disease (especially infections) and/or lack of care;

— those caused by excessive or unbalanced intake of food or of particular dietary constituents.

The prevention of malnutrition in both categories is greatly assisted if the people affected have accurate information on what constitutes a healthy diet and how they may best meet their nutritional needs. Education at all levels is important in the promotion of healthy diets and lifestyles. For those who have poor diets or nutritional problems, nutrition education and health education are strategies to influence behaviour change. Behaviour change requires motivation and efforts to recognize personal preferences, lifestyles and perhaps time constraints.

Healthy Lifestyles

Almost all governments in Asia, Africa and Latin America are advocating and working to enhance and improve development, and many international, bilateral and nongovernmental organizations (NGOs) are assisting with development in general or with specific development projects. Development involves change: cultural, social, economic and political change, and even changes in values. Any individual or group suggesting or implementing changes should consider carefully whether the outcome will be better for those affected by the change. Too often programmes and actions introduced from the outside foster change for the sake of change, or individuals or countries try to promote change to make others more like themselves, or agencies implement projects that involve change without considering the implications in terms of quality of life and with a naive assumption that all new structures are automatically better than the old ones.

Eight strategies for promoting appropriate diets and healthy lifestyles are given below. Some of these strategies do suggest changes. Where malnutrition is rampant and infectious diseases are prevalent, where these result from widespread food insecurity and a very unsanitary environment, and where the people (particularly women) lack knowledge regarding appropriate child feeding and do not understand the germ concept of disease, then clearly change is necessary if nutrition and health are to improve. There is a need for improved knowledge, improved resources and better standards of living.

In some groups of the population in the non-industrialized countries very rapid change has already taken place in the last 50 years: lifestyles have changed, age-old social practices are disappearing, and Western diets and modern ways are replacing traditional ones. Some of these changes have contributed to improvements in health, improved infant mortality rates and a reduction in certain forms of serious malnutrition such as xerophthalmia; but not infrequently these changes have also led to a new set of nutrition and health problems and to a less caring society. A steep rise in diet-related non-communicable diseases such as arteriosclerotic heart disease, obesity, certain cancers, stroke, dental caries, diabetes and others is occurring in many developing countries. Some of these problems have resulted from changing lifestyles including changed diets. Parallel with these changes there has also often been an increase in the prevalence of abandoned children, delinquent youths, child prostitution, elderly sick people not receiving proper care, and mental illness.

Not all change and not all westernization is for the good. Many poor societies possess social values that are superior to those found in many modern Western societies. Examples include emphasis on the extended family, better treatment of the elderly and infirm at home rather than in institutions, greater tolerance of the insane and more community spirit. Of course it is dangerous to glamorize life in the villages of developing countries. For many poor people life is extremely difficult; much of their day is spent doing hard manual labour, and they may lack sufficient food, housing or health care. There is no doubt that good health, a variety of social activities and, of course, enough food to eat are needed by all people everywhere. The argument here is not to oppose modernization or development, but rather to recognize, first, that all modernization and development efforts do not automatically provide benefits to the poor; and second, that some of those actions thought to be benevolent may actually downgrade the quality of life of poor people.

Adoption of so-called modern habits and lifestyles sometimes has mixed blessings. Transfer and application of modern food production technologies and food preservation and processing practices have resulted in better quality, more variety and greater safety of food available for consumption. At the same time, adoption of certain food habits and behaviours such as overconsumption of saturated fats, decline in breastfeeding and concomitant increase in bottle-feeding and cigarette smoking may be detrimental to good health and nutrition. It is therefore necessary that the potential ill effects of undesirable practices be offset by taking suitable preventive measures.

It is not suggested that change is necessarily bad. Change is inevitable and is necessary for the improvement of nutrition and health. Modern knowledge can be harnessed for the good of the poor, and each country should freely choose its actions. When change is encouraged, however, either by outsiders or by governments, it is important to consider the possible adverse effects of the changes. The question everyone should ask is: "Will the change improve the quality of life of most affected people?" Perhaps a nutrition and health impact statement should be required of all new projects before implementation, in the same way that environmental impact statements are now required in the United States.

As non-industrialized countries plan for the beginning of a new century, special attention is needed to prevent the adoption of lifestyles and dietary patterns that will lead their people into epidemics of heart disease, lung cancer, stroke, obesity, diabetes and other chronic diseases. Countries in their impatience for modernization should not neglect protecting those aspects of traditional lifestyles that are conducive to good health and nutrition A high priority should be given to protecting good traditional eating habits and good national diets; protecting good caring practices for children, the sick and the old; and protecting good moral, social and religious values The rush to modernity and to westernization could pose a major health and nutritional threat to the populations of developing countries.

Healthy lifestyles are implicit in the strategies described below, whether they refer to dietary guidelines or goals which should ensure a balanced, healthy diet or to areas such as training, education, extension or communications conducted by ministries of agriculture, education, health, women's affairs, community development, etc; workers in these strategies should be trained and employed to promote healthy lifestyles and better diets In all cases the aim should be to reduce undernutrition and infections and also to prevent the risks of non-communicable chronic diseases and health problems associated with inappropriate diets and lifestyles

It is clear that

— almost every society has a cuisine that with proper selection and little change can provide good balanced diets and that accordingly deserves protection;

— immediate action to prevent nonsmokers, especially young people and women, from becoming addicted to tobacco or from becoming users of cigarettes is vital;

— traditional good caring practices for children deserve protection and support;

— the traditional family, and sometimes the extended family, often provides a lifestyle conducive to the well-being of children;

— priority needs to be given to avoiding risky behaviours that may lead to infection with acquired immunodeficiency syndrome (AIDS) or with other sexually transmitted diseases;

— families and communities need to provide a support system for children, the sick and the elderly and to improve activity levels of those who lead a sedentary life through active recreation and sports

Lifestyles can be improved in many countries, especially for the poor, with:

— attention to hygiene, including food hygiene;

— improved sanitation and disposal of human excreta and garbage;

— safer and more plentiful water supplies;

— greater knowledge and awareness of health risks and avoidance of risky health behaviours;

— improved health services, including health care and health measures;

— agricultural improvements in rural areas, including land reform in some countries, elimination of sharecropping, better access to credit, improvements in animal husbandry and greater availability of agricultural inputs such as fertilizers, irrigation and tools

Cutting across many of these areas, lifestyles of the poor would improve if there were more equity, and those of women and children would improve if there were no discrimination against females and if there were more empowerment of women

Strategies to Influence Behaviour for Improved Nutrition

Several strategies besides education are available and have been used to influence behaviour change to improve nutrition.

Eight of them are discussed here:

— dietary guidelines and food goals,

— food and nutrition labelling,

— food advertising,

— institutional meals,

— food industry involvement,

— ensuring a consistent message,

— protecting traditional diets,
— nutrition training.

Food and Nutrition Labelling

Literate people who are interested in selecting a nutritious diet can be greatly assisted by clear and accurate labelling on food products. Food labelling that gives information on nutrient content has been used more in industrialized than in developing countries. It can be useful in almost all countries and is particularly helpful if used with a set of dietary guidelines. Other useful information on the label may include an expiry date.

The FAO/WHO Codex Alimentarius Commission has produced guidelines on nutrition labelling which deserve serious consideration by governments, especially those that do not have nutrient labelling regulations or that are dissatisfied with their existing situation. These Codex guidelines deal with pre-packaged foods and foods for catering purposes.

Nutrition labelling is often criticized for being too detailed and therefore too difficult to use. It is true that labels often list the content of some vitamins and minerals that are not causes of serious deficiencies and are not of public health importance in the country where the product is consumed. In addition to data on the nutrient content of the food and perhaps the percentages of Recommended Dietary Allowances, food labels sometimes also provide other nutritional information, for example, dietary claims such as "cholesterol-free", "low-calorie", "high-fibre" or "sugar-free". Countries need to examine these claims, to determine their accuracy and perhaps to evaluate their effectiveness. It may be more important to draw up enforceable criteria for nutritional claims. Countries moving to develop guidelines or regulations for food labelling would be wise to consult the FAO/WHO Codex Alimentarius Commission and its publications.

Food Advertising

Commercial advertising can serve to promote healthy eating, but it can also contribute to poor diets. Advertising, including food advertising, is difficult to control. Most countries expect advertising to be truthful, and truth in advertising is a basic expectation. Concerns regarding dietary claims on food labels also apply to claims made in advertisements for products and services. Advertising, particularly television advertising, of inappropriate foods to children has been the subject of much criticism and has been discussed in many reports. Most nations have agreed to the principle of regulating the advertising of breastmilk substitutes, and many have adopted appropriate legislation. However, advertising can also have a good impact on nutrition, and the food industry has an important part to play as indicated below.

Institutional Meals

A well-balanced diet is not the only advantage of institutional feeding; the introduction of new, healthier foods and food habits can also be a result. School meals, for example, provide an ideal opportunity to introduce pupils to unfamiliar foods that are nutritious and to demonstrate to children what constitutes a well-balanced meal consistent with dietary goals and guidelines.

Food Industry Involvement

Every country has a food industry, large or small, and it always has a role in promoting and influencing the consumption of healthy diets. Clearly the main objectives of industrial companies are to market foods, to make a profit and to outsell competitors. However, this can only be achieved by responding positively to public demand for particular foods. For example, the dairy industry in many developed countries has responded to the desire of people to reduce their fat and energy intakes by marketing more low-fat milk and less whole-fat milk. In general, this modification has been helpful nutritionally and has come about as consumers have become more nutritionally informed. However, changes that are beneficial to nutrition or health in the industrialized countries of the North may not be helpful in poor nations of the South. For example, where undernutrition and protein-energy malnutrition (PEM) are common and where mean intakes of fat in children are below 10 percent of total energy, a campaign to promote low-fat milk would be inappropriate.

Ensuring a Consistent Message

Nutrition education makes much more sense to the public if there is some degree of consistency in the main messages. This is not to suggest that there is a need for control or censorship; but regarding nutrition and health, people are often confused because they hear different and sometimes conflicting messages. For example, many extension workers and others in ministries of agriculture may be promoting the consumption of diversified, energy- and nutrient-dense diets as the way to overcome micronutrient malnutrition, while others may be undermining these messages by advocating the widespread distribution of dietary supplements in the form of pills and capsules.

If nutrition educators can agree on the main nutritional problems and then on the appropriate advice to provide to the public, everyone's work becomes easier.

Consistency is important in all aspects of information, not just in content. Nutrition education must not distinguish four food groups according to one ministry and three according to another. Similarly, national agricultural and food policies need to address the nutrition problems of the country, and the ministry of health needs to promote sustainable solutions to the major deficiencies by advocating food-based approaches that address the fundamental problems of widespread poverty and food insecurity.

Protecting Traditional Diets

A neglected but important topic, protection of dietary tradition is especially relevant for those countries where diet-related chronic diseases are not prevalent but where economic development permits at least some people to purchase a wide variety of foods, including animal food products.

In general, traditional diets in Asia, Africa and Latin America are based on cereals or root crops, with significant amounts of legumes, fruits and vegetables. Often poultry, meat and dairy products provide only a small proportion of total energy but are appreciated as side dishes or tasty additions to the staple foods.

Relatively low in total fat, saturated fat and cholesterol, these diets are high in complex carbohydrates and fibre. If plenty of fresh vegetables and fruits are also consumed, these diets are often rich in carotene and vitamin C, which are antioxidants.

Protecting good traditional diets starts with protecting or enhancing the production and marketing of traditional foods. Working with the local food industry to help in safe food preservation and packaging is important, and making foods easier to prepare for the table would contribute much to their popularity. One obvious attraction of many Western dishes is their convenience; busy people are attracted to them and homemakers can save time using them.

Nutrition Training

Most countries have far too few professionals who are expert in or knowledgeable about nutrition. Moreover, nutrition training is often a neglected topic for persons other than nutritionists and dietitians. A wide range of professionals could benefit from more and better training in nutrition: health professionals such as doctors, nurses, midwives, health assistants or auxiliaries; agricultural staff, including extension workers, research scientists and high-level ministry officials; teachers and others throughout the formal and non-formal education system; social and community development workers; workers in institutional feeding; staff in NGOs involved in development, health, agriculture, community development and other activities; professionals in the food and related industries; and many others.

A prerequisite for designing appropriate training at suitable levels is a review of the nutrition content of the curricula of training institutes of many kinds in various fields such as health, education, agriculture and community development. Most institutes will be found to have inadequate coverage of nutrition. If this proves to be the case, a group of experienced persons might be formed to make recommendations regarding strategies for improving nutrition training, changes in the curricula and the means to make the changes.

The first need might be to train the trainers. In poor countries this may require external assistance. In designing the training several questions need to be addressed. What are the most important topics in training, taking account of the most important nutrition problems? What do those being trained need to know in order to integrate nutrition in their jobs? Can some progress be made in the near future with the organization of short courses?

Nutrition Education and Communication

Nutrition education is a strategy that has been widely used for many years to promote healthy diets and thereby ensure proper growth of children and a reduction in all forms of malnutrition. The basis of any nutrition education programme should be to encourage the consumption of a nutritionally adequate diet, to promote healthy lifestyles and to stimulate effective demand for appropriate foods.

In the past, nutrition education was too often conducted in an unimaginative way. People were instructed to eat this or that food because it was "good for you". Attempts were sometimes directed at making radical rather than gradual changes in the diets of the people who were the targets of the nutrition education. As a result, very few of the nutrition education programmes were successful.` They were frequently carried out by persons of a different culture or social class from those being educated. The lessons of history show clearly that nutrition educators should start from the premise that most mothers are doing their best to feed their families properly. If they are not managing to do so, the reasons may well be beyond their control.

In most circumstances the nutrition education content must be formulated on the basis of a problem analysis. The education must be relevant to the reality.

Inadequate total intake of food by young children (energy deficiency) is the main cause of malnutrition in Africa, Asia and Latin America. Therefore, initial advice might be to feed a malnourished infant with the same food as before but more frequently, or to provide just a little more of the food. This advice should be more acceptable to parents than an attempt to make major, often unrealistic, changes in the diet. Other recommendations for change should be simple and feasible for the family, consistent with its cultural habits and of course nutritionally sound.

Nutrition education has frequently failed because the advice did not conform with the above criteria. Throughout the world there have been examples of nutrition education messages that urged poor mothers to provide their children with meat or fish every day, or one egg per day or three cups of milk per day. This advice may have been nutritionally reasonable, but in all other respects it lacked sense. Except in very few communities and countries, poor families cannot afford to provide these foods to their young children so often, and it is now known that it is unnecessary to do so.

Nationally, nutrition education may be carried out by several ministries (health, agriculture, education, social or community development, etc.) and also by various NGOs. All these bodies should agree on common objectives for a nutrition education programme, and each ministry must decide how it plans to implement it. Factors that should be decided upon, which are rarely clearly defined, include the content of the message (discussed above), the target audience for the programme and the media to be used. This strategy may appear simple, but its application will require a change in both the philosophy and the operation of most nutrition education programmes.

The choice of media depends on the formal and informal information and communications infrastructure of the area in question. In general it is wise to use a combination of communications media in an integrated way. However, a concentrated radio campaign may often be the cheapest and most effective way of reaching the bulk of the population. In addition to stations controlled by the government, commercial radio and television should be used for nutrition education. Certain priority issues or areas of concern should receive concentrated effort.

As mentioned, the stress should be on small changes that will complement existing dietary practices and not on major changes. Failure has occurred in past campaigns that attempted to impart a mass of general information on nutrition, rather than hitting hard with a few well-designed messages in a limited number of priority areas.

The efforts of the various ministries and organizations involved in nutrition education should be closely coordinated so that the messages received from different sources will complement and reinforce each other.

Who should give nutrition education? When should it be offered? To whom should it be aimed? The answers to these questions are in general quite simple. Everyone who has the knowledge (for example, members of health teams, schoolteachers, agricultural extension workers) should provide nutrition education. They should do so at every possible opportunity (for example, the doctor when treating a patient, the midwife at the antenatal clinic, the health nurse when visiting a home, the extension officer at a farmers' meeting, the schoolteacher in a class or at a parents' meeting). Every person in the country should be the target of nutrition education. Even if the message concerns PEM in the preschool child, for example, the problem is so important that all people can benefit from being informed about it.

Perhaps the most persistent and frequent error that has been made in nutrition education has been the overriding attention given to animal protein. It is now generally agreed that protein deficiency is not the main dietary shortcoming to be overcome, and that even if it were, animal products do not offer a reasonable or feasible solution in most poor societies. PEM, which is the most important nutritional problem, is much more often the result of a low total intake of food by the child, who may then be deficient in both energy and protein. The solution is to increase the quantity of foods already eaten. If efforts are to be made to increase protein intake, then the stress should be on vegetable foods that are rich in protein, such as legumes, rather than on animal products. Nevertheless, in many nutrition education programmes of the past 40 years emphasis has been placed on increasing the consumption of meat, fish, milk, eggs and manufactured protein-rich foods. This education has totally failed because economic reasons have precluded the adoption of the advice and frequently the foods recommended have not been easily available.

Nutrition educators have much more to learn from commercial advertising, which has often been successful in changing food habits and attitudes. Commercial promotion uses the media skilfully. The talent available in commerce should be harnessed more often to assist with nutrition and health education.

Past nutrition education initiatives have had some successes but many failures in terms of improving dietary intake and reducing the extent of malnutrition in a community or a country. The failures have occurred not mainly because nutrition education is a wrong strategy, but rather because the methods used have not led to the desired behaviour change.

In the past 30 years new approaches have been used to elicit changes in behaviour with a nutrition objective, and there is evidence that some have been more successful than older, more traditional approaches. One approach, which has been termed "social marketing", uses some principles from commercial marketing. Other approaches using principles adopted from the behavioural sciences have also improved nutrition education efforts: nutrition educators seek to identify the nutrition problems and eating behaviours of people within the social context in which they live, taking cognizance of cultural factors; only then are the communication techniques chosen and appropriate messages formulated for specific or general audiences.

Social Marketing

In recent years social marketing to promote improved health and better nutrition has been widely, and sometimes successfully, used. There have been some real success stories.

A major difference between traditional nutrition education and the newer social marketing approach is that the latter starts with what commercially would be called "consumer research". An attempt is made to discover, using various techniques such as surveys and focus group interviews, what the consumer or the public is doing and why. The older nutrition education approach started from the premise that malnutrition exists, that people's diets are bad and that people need to be told to eat a good diet providing foods in all four basic food groups. The new approach would be likely first to use consumer research to identify a few important problems such as decline in breastfeeding, infrequent meals or drinking of contaminated water, and would then address them. The results of the research regarding consumer views, perspectives and practices lead to decisions on appropriate messages, communication techniques and targeting.

In the commercial world test-marketing is nearly always done before launching a product. It may also be sensible in nutrition education using social marketing and modern communication techniques; the messages developed and the communication techniques chosen to tackle the problems assessed and analysed may be tried out in a limited way. They can then be reassessed, reanalysed and modified, changed or abandoned, before they are implemented for a larger audience.

These methods, if successful, could lead to a major national campaign or to more limited nutrition education activities in certain communities; they could lead to use of television time or to the use of communicators in the villages. The major difference between these and older methods is a recognition that people have reasons for their behaviour and that nutritionists need to respect and learn from people before they attempt to change their behaviour. It may be necessary to identify the resistance points that impede change. Any nutrition education approach that incorporates empowerment and respect for local culture is more likely to be successful than those that do not.

FAO and other United Nations organizations provide assistance for the development of appropriate nutrition education programmes. They maintain that nutrition education should be carried out broadly through schools, newspapers, television, radio and other mass media as well as through face-to-face contact.

Preventing Specific Micronutrient Deficiencies

Over 30 micronutrients are essential for human health and for the proper growth and development of children. They are all vitamins and minerals available in foods. Micronutrient deficiencies are prevalent public health problems in many countries, especially developing countries. The micronutrient deficiencies that are most prevalent in the world are those of vitamin A, iodine and iron. Together with protein-energy malnutrition (PEM), these deficiencies constitute the "big four" nutritional problems. There is wide geographic variation in their prevalence.

In the early 1990s almost all countries pledged to devote major efforts to eliminating vitamin A and iodine deficiencies and substantially reducing iron deficiency by the year 2000. These tasks will be more difficult for some countries than for others, but all countries where these micronutrient deficiencies exist should have a policy and strategies to deal with them. However, the initiatives should not undermine, replace or reduce efforts to control PEM, which is often more prevalent and more important as a public health problem. In some countries some other micronutrient deficiencies may also constitute a public health problem and may perhaps be more important than the deficiencies of vitamin A, iodine or iron. In these countries appropriate attention needs to be devoted to the most important deficiencies based on their prevalence, the extent of the morbidity they cause, their contribution to mortality rates, their social and public health significance and, finally, the feasibility and cost of control.

Individual countries and communities can take many different strategies and actions to address these micronutrient deficiencies. It is important to make certain that the strategies and actions are coordinated and that consideration is given to strategic actions that address more than one nutrition problem simultaneously.

Comprehensive Versus Targeted Approaches

Policies and programmes designed to control the three major micronutrient deficiencies are usually either comprehensive or targeted. A relatively comprehensive (or holistic) approach to deal with vitamin A deficiency might include public health measures, horticultural activities, treatment and control of infections, fortification of foods and judicious use of vitamin A supplements, allied with government activities to reduce poverty and improve food security. A narrowly targeted approach might be a distribution of high-dose vitamin A capsules to young children at high risk of vitamin A deficiency.

The comprehensive approach can be compared to firing a shotgun: many small pellets are fired, rather than a single bullet, and these may hit a wider area or different targets. The targeted approach, on the other hand, is analogous to using a rifle: there is one bullet, which is lethal, but only if it hits the target. Thus it has sometimes been termed the "magic bullet" approach.

For many public health problems and most types of malnutrition, the holistic approach is philosophically and politically preferable and is more likely to be sustainable than the narrow, targeted approach. In practice, the place for the magic bullet is in dealing with a single problem or individual.

Holistic approaches might appear to be more daunting, more difficult and perhaps slower for reaching the optimistic goals for micronutrient deficiency control. However, this need not be so, because the holistic approach can also embrace the magic bullet approach. A vitamin A deficiency control strategy, for example, can include targeting of high-dose vitamin A supplements along with initiatives for increasing production and consumption of carotene-rich foods, fortification, nutrition education and broad public health measures. Optimism that holistic approaches will be successful depends to some extent on a favourable political and social climate and some chance of social mobilization and community participation. Favourable economic development is a helpful but not necessary condition.

The goals of virtually eliminating vitamin A and iodine deficiencies and reducing iron deficiency substantially by the year 2000 are ambitious, but they are realizable in several countries. In all cases their realization will require a rather rapid and sustained increase in levels of appropriate activity. Achievement of the goals will depend not, as is often stated, primarily on political will, but more on government actions. Will is important, but actions are essential. Many international agencies, non-governmental organizations (NGOs) and others are poised to assist countries and their local experts in concentrating efforts to control micronutrient deficiencies. FAO, the United Nations Children's Fund (UNICEF) and the World Health Organization (WHO) are among the agencies concerned.

A Micronutrient Deficiency Control Plan

The first requirement, which some countries have already met, is to formulate a national plan with defined strategies and actions and clear lines of authority to take action. In most cases an overall micronutrient plan is desirable. However, specific deficiencies may call for different control strategies, involving different professionals and perhaps necessitating separate plans of action.

The prevalence of each deficiency in different parts of the country and the underlying determinants may or may not be well known. Action should not await new comprehensive nutritional surveys, but more detailed assessment of the micronutrient deficiencies and their underlying causes may be desirable. This can also provide baseline information to judge the effectiveness of actions taken. Baseline information on the prevalence of the deficiencies is often usefully supplemented with specific information on food intakes; relevant social, cultural and economic factors; and data on the health situation.

Four Control Strategies

Four main strategies can be implemented to reduce or control micronutrient deficiencies. They operate in concert with broader strategies to improve the quality of life in particular countries and communities. Actions at all levels - international, local and family - to improve household food security, individual health and care can have an impact on micronutrient deficiencies and should always be taken into account in micronutrient deficiency control strategies.

The four basic micronutrient strategies are:

— improving diets, especially dietary diversity;

— public health actions;

— fortification or nutrification of foods;

— providing medicinal supplements.

These four strategies are listed in order of sustainability; clearly improved diets contribute to controlling a micronutrient deficiency in a much more sustainable way than medicinal supplements. Public health actions and fortification are of intermediate sustainability. Many public health measures, such as improved health knowledge, water supplies and hygiene, remain in place, whereas other measures, such as immunizations, require continuing action. Undoubtedly conferring the knowledge and ability to produce, procure and consume an appropriate diet is the most sustainable way to prevent micronutrient deficiencies.

Improving Diets

Clearly the ultimate goal in attainment of micronutrient food security is to ensure that people consume a diversity of foods that provide them with the required quantities of all essential micronutrients on a continuing basis. This surely should be the basic long-term strategy of all governments addressing the problems of vitamin A and iron deficiencies. For infants, the protection, support and promotion of breastfeeding and emphasis on the health and good nourishment of the mother offer the best protection. To prevent iron and vitamin A deficiencies in adults, stimulating the production and consumption of micronutrient-rich foods is vital.

Nutrition education is an important part of this strategy. However, it will be effective only if the appropriate foods are available. Education to improve production and especially consumption of appropriate micronutrient-rich foods must go beyond old nutrition education methods which exhorted people to consume certain foods because they were "good"; education programmes must be designed to elicit behaviour change that will be permanent. A programme in Thailand, for example, successfully used social marketing methods to raise dietary vitamin A intakes in the northeastern part of the country, and Bangladesh has seen some successes in increasing home or village production and consumption of carotene-rich foods.

Improving dietary diversity is best considered as an integral part of community actions to improve household, and then child, food security. The actions planned will often be cooperative and may include agricultural activities, school-based projects and assistance to families, both urban and rural.

This sustainable approach to the control of micronutrient deficiencies is often criticized as being too difficult or at best a very long-term strategy. Recent examples from many parts of the world, however, suggest that good results can be seen in a relatively short time. Critics of this strategy are often those who are philosophically tied to "quick-fix", medically oriented solutions that can be planned from outside the country or outside the community. But the food-based strategy is sustainable and is the only one that controls vitamin A deficiency permanently.

Public Health Actions

Clearly any measures that reduce infections and promote good health will also help to reduce most micronutrient deficiencies, especially vitamin A and iron deficiencies.

Specific health actions in the control of micronutrient deficiencies include early diagnosis and treatment of deficiencies. When a deficiency is recognized early and properly treated it cannot lead to serious consequences. Thus recognition by health workers that preschool-age children in a community have night blindness or Bitot's spots, that schoolchildren have small enlargements of the thyroid gland or that women have low haemoglobin levels can prompt timely medical action and cure. This evaluation can be part of primary health care.

At the next level are public health actions, particularly those that control infections. They include immunizations against infectious diseases; mass deworming and measures to reduce the transmission of parasitic infections; and improvements in sanitation, household hygiene and availability of safe potable water. All can help in the control of micronutrient deficiencies. Good maternal and child health services, availability of family planning, health and nutrition education and household and environmental hygiene measures contribute to reducing malnutrition.

Some of these health interventions are highly sustainable and many will have an impact on nutrition and health beyond the micronutrient deficiencies.

Fortification or Nutrification of Foods

Food fortification, usually salt iodization, is widely recognized as the most important strategy for the control of iodine deficiency disorders (IDD). Fortification can also contribute to the control of vitamin A and iron deficiencies among populations who purchase food and can afford fortified ingredients. Many different foods in industrialized countries are fortified with iron and vitamin A. Many North Americans get more than their total daily requirement of vitamin A and iron from one large bowl of a fortified breakfast cereal and from a slice of toast liberally spread with margarine fortified with both carotene and vitamin A. It is believed that food fortification was responsible for the control and often the virtual elimination of many serious micronutrient deficiencies that were prevalent in industrialized countries early in the twentieth century.

Food fortification has to be continued as long as there is a risk of people suffering from a particular micronutrient deficiency and dietary diversification or other steps are not removing the risk. The sustainability of a fortification programme depends on food industry cooperation, monitoring and legal enforcement.

Fortification, including salt iodization, has been successfully used for many years in industrialized countries, but in some developing countries there have been serious problems in introducing it. A national programme requires advocacy, political will and often multisectoral actions or involvement, with the participation of several ministries. It also requires cooperation from the food industry, whose opposition would make fortification difficult if not impossible. An early step to successful implementation is often the establishment of an interdisciplinary committee, including people from universities or research institutes who have conducted research on the problem; representatives of appropriate ministries including health, commerce and industry, finance and perhaps education and agriculture; and representatives of the food industry. Consideration can be given to fortifying more than one commonly eaten food.

Medicinal Supplements

The provision of micronutrients taken orally or by injection is usually simply called "supplementation" rather than "medicinal supplementation", but in fact these supplements are generally provided as medicine or used in a medicinal sense.

The major role of supplementation with iodine, vitamin A or iron is as a short-term measure. It may be used in the longer term for individuals at special risk of the deficiency. Programmes of medicinal supplementation should usually be introduced for rapid improvement while long-term, sustainable interventions are planned and readied for implementation.

In some instances medicinal supplements may be the only feasible intervention to protect people. They are especially useful in the event of natural or civil disasters, when no alternative strategy may be immediately available.

Medicinal supplementation is the least sustainable strategy because it depends, first, on a delivery system that reaches almost all persons at risk of the deficiency and, second, on active participation, including behaviour change, by those at risk of the deficiency (or in the case of the children, by their families and guardians). These two essential components are very seldom fully realizable, and this is one reason for failure.

However, as indicated below, there is a good distance between a rejection of all supplementation and a decision to attempt a national programme to provide a medicinal micronutrient supplement (such as high-dose vitamin A capsules) to all children between six months and five years of age throughout the country. The middle ground is the usual choice and is most appropriate; this includes medicinal supplements for persons at special risk, as well as broader programmes, for example, provision of oral iodine to non-pregnant females of child-bearing age in IDD-endemic areas to protect their future foetuses from iodine deficiency while salt iodization is being introduced.

Micronutrient supplementation is more effective if it reaches people through existing delivery systems, for example, when iron is routinely given in antenatal clinics, vitamin A to malnourished children when they come for growth monitoring and oral iodine at school to female pupils 14 to 19 years of age. It has been suggested that high doses of vitamin A be given to infants at the time of immunization as part of WHO's Expanded Programme on Immunization (EPI), but this proposal should probably not be recommended. The infants would be "captive" subjects, but infants in their first six months of life are usually breastfed and at low risk of xerophthalmia, and there is evidence that high doses of vitamin A in young infants can cause undesirable reactions. Similarly, more and more projects have been aimed at providing schoolchildren perhaps once a year with an anthelmintic drug to rid them of intestinal worms, and vitamin A promoters have suggested that high doses of vitamin A be given at the time of deworming. However, school-age children usually do not have serious clinical manifestations of vitamin A deficiency. Targeted use of micronutrient supplements should aim at people at special risk of the deficiency, not at people who are easy to reach but who have little risk of the deficiency.

Preventing Vitamin A Deficiency

The reduction and eventual control of vitamin A deficiency in most poor countries where it is prevalent almost always requires a broad approach. Seldom will it be appropriate to use only one strategy.

The United Republic of Tanzania is one of several countries taking a broad approach. The country's interdisciplinary, interministerial national micronutrient committees have set actions in place to improve dietary intakes of vitamin A-rich foods. They include horticultural activities and nutrition education; public health actions of various kinds; an exploration of possible foods to fortify; and judicious use of high-dose vitamin A supplements, widely available through health services. At the same time, Tanzania is striving, by means of economic, agricultural and other policies, to improve the well-being of poor Tanzanians in a sustainable way, which if successful will also work to reduce vitamin A deficiency.

Each country needs to consider to what extent it will aim to use each of the four possible strategies described above. Communities and families also take their own actions, becoming participants to a greater or lesser extent in strategies planned nationally.

Improving the Vitamin A Intakes of At-risk People

In developing countries most people get most of their vitamin A from carotene in foods, not from preformed vitamin A, which is present only in foods of animal origin. Therefore endeavours to increase dietary diversity to improve intakes of vitamin A will focus mainly on raising the intakes of foods containing carotene. Certainly there is some place, if appropriate in view of incomes and availability, for modest attempts to increase intakes of foods of animal origin that contain vitamin A, but the main step is to promote the consumption of carotene-rich fruits and vegetables. Other sources of carotene in certain countries are red palm oil and yellow maize. It is also important that diets contain adequate fat, which assists with absorption of carotene, and enough protein for retinol transport.

To increase intakes of vitamin A- and carotene-containing foods, including breastmilk, it will often be necessary to stimulate changes, first in production and availability of these foods and second in consumption, especially by those who are at risk of vitamin A deficiency.

Several projects have led to improvements in knowledge, attitudes and practice relating to consumption of vitamin A-containing foods and in some cases to improvements in vitamin A nutritional status. In Thailand and Indonesia social marketing and other methods were successfully used to increase consumption of vitamin A-rich foods. In Bangladesh the emphasis was on home production of carotene-containing foods and the consumption by children of more green leafy vegetables and carotene-rich fruits. This project was allied with an endeavour to raise families' awareness of night blindness as a sign of vitamin A deficiency. Reductions in night blindness then illustrated the success of the project. In the Philippines and Indonesia projects in selected communities have attempted to increase children's complementary consumption of foods rich in vitamin A with foods containing adequate fat. A dietary approach in the United Republic of Tanzania involves a broad set of activities, including information, education and communication components aimed at creating public awareness of the vitamin A problem and stimulating increased production and consumption of vitamin A-rich foods. There is wide use of radio and newspapers. Special efforts are being undertaken to improve horticultural practices and to link these with control of vitamin A deficiency. Work is under way to increase the production and improve the marketing of red palm oil.

Breastfeeding is protective against vitamin A deficiency. Colostrum is also rich in vitamin A. The baby who is exclusively breastfed for four to six months is protected against xerophthalmia, and for babies six to 24 months of age breastmilk provides very important amounts of vitamin A. For these reasons protecting, supporting and promoting breastfeeding is a very important strategy in the control of vitamin A deficiency. Breastmilk will provide more vitamin A if the mother has an adequate intake of the vitamin. Therefore foods rich in vitamin A should be promoted not only for young children, but also for women of child-bearing age and those who are breastfeeding their babies.

At the community level the health worker, schoolteacher, extension officer or social worker needs to emphasize the importance of vitamin A-rich foods for children and pregnant and lactating women. Families need to know which local foods of those that they can afford to buy and that their children will willingly eat are rich in carotene. Often children will prefer mango, papaya, yellow sweet potato and pumpkin over green leafy vegetables. When red palm oil and liver are available, children should have priority in getting these. Families might be assisted in growing vitamin A-rich foods in urban or rural gardens and in preserving them. Another action is to inform families how to prepare vitamin A-rich foods for consumption by children). The foods prepared and served to children differ from society to society, but cooked green leaves put through a sieve or shredded with a little oil or with groundnuts, or mashed cooked pumpkin, sweet potato or carrots, will often be appropriate.

The strategy of improving production and consumption of vitamin A-rich foods is the only sustainable long-term solution for controlling vitamin A deficiency. In most countries it should be a high-priority strategy.

Public health actions

The first health-related action is to ensure that health workers, especially those who see children in out-patient and in-patient facilities within the primary health care system, easily recognize xerophthalmia and appreciate those conditions and illnesses that raise the risk of vitamin A deficiency. Having made their diagnosis or assessed the risk, they must also be in a position to provide appropriate treatment, usually a high dose of vitamin A given orally. Of particular importance is routine oral administration of high-dose vitamin A to all measles cases: 200 000 IU for children over two years of age, and half this dose for those under two.

The second health-related action is to treat, and more importantly to control, infectious diseases. Many infectious diseases exacerbate vitamin A deficiency and not infrequently push a vitamin A-deficient child into overt xerophthalmia. Prevention of measles by immunization is a vitamin A intervention. Vitamin A given to a child with measles greatly reduces the risk of death. Infections influence vitamin A status by reducing appetite, thus lowering food intake and vitamin A intake. Viral, bacterial and parasitic intestinal infections may also reduce vitamin A absorption or conversion of carotene to retinol. Infections are made worse by PEM, which almost always is present in children with xerophthalmia.

The third health-related action is to take steps to control disease and promote health, because these might influence vitamin A status. Deworming of children, treatment and control of diarrhoea and respiratory infections, immunizations and improved sanitation and water supplies can all have a role. The health sector's support of breastfeeding will also help in the control of vitamin A deficiency. Health and nutrition education also contribute. At the community level, it is important that families be motivated to have their children immunized, to seek early treatment, to control infections and to improve personal, food and household hygiene.

Fortitication with vitamin A

Fortification is an attractive strategy, especially when compared with medicinal supplementation, because the market system delivers the nutrient. When one or more commonly eaten foods are fortified with vitamin A, behaviour change is not required, and there is no need for a cadre of workers to take vitamin A capsules from house to house or for the major government expenditure that is required for supplementation. Fortification is usually a relatively low-cost intervention for governments. Once in place it needs to be maintained and perhaps mandated for the food industry by legislation. Thereafter it is a relatively sustainable intervention, unlike medicinal supplementation. Monitoring may be all that the authorities need to do.

The methodologies of vitamin A fortification are well tested. Hundreds of different foods have been fortified, mostly in industrialized countries without at-risk people especially in mind. Breakfast cereals of all types (based on maize, rice, wheat or oats), margarine, dairy products and other foods are fortified. Food technologists, who long ago developed methodologies for adding vitamin A to oils and fats, are now able to add the vitamin to many other foods. In developing countries the vehicles used for vitamin A fortification include monosodium glutamate (MSG), sugar, tea and margarine.

In the past, developing countries have tended to seek only one widely consumed food as the vehicle for vitamin A. Because the technology exists to fortify many foods, it now seems preferable to consider fortifying several foods simultaneously to achieve wider coverage. The risk of toxicity needs to be considered, especially where quality assurance is difficult to achieve. Industrialized countries such as the United States fortify many foods and do not report widespread cases of toxicity.

Fortification has not been an easy strategy to initiate and sustain in developing countries. In many countries the major problems of vitamin A deficiency are in children who may consume mainly local foods and very few foods that are centrally processed at a facility where vitamin A could be added. Another problem is the cost of the fortified foods and their affordability for the poorest, high-risk groups.

Nonetheless, national committees given responsibility for developing strategies to control micronutrient deficiencies to meet the goals of the World Summit for Children and the International Conference on Nutrition (ICN) need to give serious consideration to fortification for control of vitamin A deficiency. They may need the assistance of outside expertise, and United Nations agencies are often ready to provide it, but local food scientists and food technologists should be brought into the effort and should begin investigating the possibilities of fortification as a strategy. Thereafter it is necessary to consider the foods widely consumed by the poor that could be fortified. Other decisions before a trial is undertaken include consideration of what form of vitamin A to use and at what levels, the cost and how and where the trial should be conducted. After the trial, it is necessary to consider whether legislation is needed, how monitoring will be conducted, how quality control will be'ensured and who will bear the costs.

Often, if all of a particular food is to be fortified, consumers can bear the cost: if all sugar or all MSG sold in a country is fortified with vitamin A, the price of the product can be raised very slightly per amount purchased. This is usually the best option. In a trial in the Philippines vitamin A and a flow-enhancing substance were added to MSG. The public usually purchased 2.4-g packets of MSG to add to soups, stews or other foods. It was decided to add 0.1 g of the fortificant and to reduce the amount of MSG per packet to 2.3 g to maintain the same packet weight. As the MSG costs more than the fortificant, the packet could thus be sold at the old price. It does no harm if families consume very slightly less salt, sugar or MSG per day.

Fortification of foods with vitamin A has been difficult in several countries mainly because of political constraints or industrial opposition and sometimes because of misinformation by advocates opposed to the use of the food vehicle or to the principle of fortification. Fluoridation of water supplies has received similar opposition.When vitamin A fortification is implemented consideration might be given to simultaneous fortification of the chosen foods with iron and perhaps other micronutrients.

Medicinal vitamin A supplements

Vitamin A is a fat-soluble vitamin; once absorbed, it is excreted slowly and a good proportion of a high dose remains for some time in the body. Therefore large doses of vitamin A can be given at long intervals.

About 30 years ago it was found that 200 000 IU of vitamin A given to children aged one to five years protects them from vitamin A deficiency for some weeks. Most programmes provide vitamin A every six months, but by then serum vitamin A levels may have returned to deficient levels, so dosing every four months is probably preferable.

Governments using medicinal vitamin A supplementation sometimes attempt universal supplementation to reach all children of a defined age group in the country or perhaps in certain regions of the country. However, this approach has usually failed to reach the objectives set, has proved expensive, has required a complex delivery system, has had coverage rates that have dropped off rapidly after the first dosing and has missed the children most at risk of xerophthalmia. Populous countries with serious vitamin A problems such as India, Indonesia and Bangladesh have attempted nearly universal supplementation at least in some regions of the country. These programmes have no doubt benefited some children, but in general continuing attempts at universal supplementation are not justified. In Indonesia, the major reduction in xerophthalmia undoubtedly resulted more from general improvements in the standard of living of poor people, better household and national food security, improved health services, general improvements in the economy and national attention to nutritional problems than from high-dose medicinal supplements. Major reductions in infant and young child mortality and lowered rates of nutritional marasmus occurred simultaneously with the reduction of xerophthalmia.

Many countries are now targeting vitamin A supplements to particular groups or, more commonly, making them available to those at risk of vitamin A deficiency when they come in contact with the health care system. Free or subsidized supplements are made available to health centres, dispensaries, clinics and hospitals. This strategy has advantages over universal supplementation.

Groups to be targeted for supplementation might include all cases with signs of xerophthalmia, measles, moderate or severe PEM, gastro-enteritis or other selected diseases and conditions. In some countries vitamin A supplements have been tied to other health interventions, for example, child immunizations. This approach should probably be confined to children over six months of age. Supplementation could be combined with regular deworming of young children and with growth monitoring for children who have poor growth. It is also important to give supplements to children in refugee camps or in times of drought or famine. Providing women before pregnancy with vitamin A supplements is not recommended because of the increased risk of birth defects.

When selective supplementation is first introduced it is important to follow the Tanzanian example and train the primary health care workers in the appropriate use of vitamin A supplements. One- or two-day courses, led by a travelling team of trainers, can provide simple literature (e.g. a short hand-out), educate health workers on signs of xerophthalmia and present an agreed list of conditions that warrant vitamin A doses.

In all supplementation programmes there is a need to establish a record system to reduce the possibility that children will get high-dose supplements too frequently and therefore risk toxicity. Vitamin A supplementation programmes should be used in combination with activities to improve dietary intake of foods rich in vitamin A and with public health measures aimed to reduce vitamin A deficiency. The use of fortification should also be taken into consideration.

Preventing Iodine Deficiency Disorders

Iodine is the easiest of the three important micronutrient deficiencies to control. The strategy most strongly recommended is not dietary improvement but salt fortification, usually termed salt iodization. Public health measures are not an important strategy for the control of IDD, but medicinal supplementation can have a place in highly endemic areas, especially as a short-term measure while salt iodization is being introduced. Iodine is an absolutely vital nutrient, but humans require it in tiny amounts. Adults should consume 100 to 200 μg of iodine per day; this amounts to less than a spoonful of iodine per person every 50 years.

Improving diets

Nutrition education and other methods to influence people to change their diets do not work as measures to control IDD because the iodine content of foods depends more on geography than on the foods. The iodine content of plants is much affected by the iodine content of the soil in which they are grown. Thus most foods grown in soils depleted of iodine, found most frequently in highland areas, are deficient in iodine. The vegetables, cereal grains, legumes and other foods grown in iodine-depleted soils high in the Andes or Himalayas have much less iodine than those grown in the lowlands near the mouth of the Amazon River or in the Ganges Delta. Influencing higher consumption of particular local foods is therefore not effective. Seafood and seaweed are rich sources of iodine, because sea water has high levels of the mineral. However, these foods cannot be promoted in areas far from the sea.

Nutrition education and other methods to influence behaviour change can be used to reduce consumption of foods containing goitrogens, such as cabbage and other vegetables of the genus Brassica and also some kinds of cassava. In countries where salt is available in both iodized and non-iodized forms, nutrition education and other means should be used to encourage people at risk to use the iodized salt. Nutrition education can also serve to explain the cause of the problem and to stimulate demand for government and other action.

Public health actions

No specific public health measures are used to control IDD. However, good health care and medical services are useful in the diagnosis of goitre and hypothyroidism and in the recognition of cretinism and of metabolic and neurological problems in children whose mothers were iodine deficient during pregnancy. Large nodular goitres that do not respond to iodine or other medical therapy may require surgical excision.

Fortification

It is almost unanimously agreed that fortification is the most effective strategy for the control of IDD. Iodine has successfully been added to water, bread, milk, various sauces and mixed foods, salt and other foods. Recently research has again focused on adding iodine to drinking-water as a means of controlling IDD, but iodizing salt is the major recommended strategy to control IDD by the year 2000.

In temperate climates potassium iodide has been most widely used, but in tropical countries potassium iodate is recommended. It mixes readily with salt at levels from 40 to 100 mg of iodine per kilogram of salt. It is more stable and less likely to be adversely affected by heat and humidity than potassium iodide. The level of fortification varies from country to country and should be based on two considerations: mean levels of salt intake by at-risk populations and other sources of iodine in the diet.

The technology for iodine fortification of salt has been known for a long time, and it is a simple, relatively inexpensive process. It does not change either the appearance (including the colour) or the taste of the salt.

It is believed that once a government manages to get the iodization of salt well established and supported by legislation, it provides by far the best solution to the control of IDD for those who consume the salt, and the control should be sustainable. Many of the industrialized countries have maintained salt iodization for decades and have controlled IDD.

For a variety of reasons, not all of which have been fully elucidated or publicized, iodization of salt in many developing countries, even when legislated, has not been successful. It has not failed because the technology is wrong, but because of other failures in the system. To work, the strategy requires not only political will, but genuine political and government action; honest and incorruptible people at all levels, from top government officials to lower-level technologists; well-trained personnel with knowledge and expertise; social support for the exercise; and finally adequate funding. Control of IDD is an intervention for which poor countries can usually quite easily get support from organizations such as FAO, UNICEF, WHO, the World Bank and bilateral aid agencies. At US$0.05 per person per year, iodization of salt is a very cheap intervention.

It should be noted that the availability of a solution that produces a colour if added to salt that contains iodine has made it much more feasible to monitor salt locally to make certain that it is iodized. This is, of course, more a qualitative than a quantitative test.

In countries where iodization has been tried but has not seemed to work or where implementation has been fraught with difficulties, it is vital to assess the problems and the resistance points. Salt is a profitable, commercially marketed product, and efforts to develop successful partnerships among governments, the salt industry, retailers and consumers can make salt iodization successful.

Medicinal iodine supplements

Iodine can be provided medicinally to cure IDD, to reduce goitre size and to prevent IDD, including cretinism. Widespread dosing with either oral or injectable iodine has been used in high-risk areas and may be a suitable strategy to reduce IDD quickly while salt iodization is being introduced. Unfortunately, often much more time passes than planned before iodized salt is generally available and consumed.

The preparation most widely available is Lipiodol, which provides 480 mg iodine in 1 ml of oil. It can be either given by injection or taken orally. The doses of iodine in oil, which are much higher than daily physiological needs, are designed to work prophylactically. They provide iodine that lasts for many months. Injections of iodized oil are claimed to prevent IDD for three to four years, and oral iodine capsules for one to two years. Good evaluations have not been done.

In children injections of iodine in oil should be given in the thigh or buttocks. In adults and older children the thigh or buttocks can be used, but the upper arm is better. Oral iodine is often provided in capsules which are swallowed or as a liquid given using a dispenser or syringe which gives a measured dose into the mouth, if possible without touching the lips or tongue.

Oral iodine has many advantages over injectable iodine. It can be given by persons who are not trained to give injections, and therefore it is cheaper to provide. More doses can be given per hour. Above all, there is no risk of spreading acquired immunodeficiency syndrome (AIDS) or other infections which can be spread by syringes and needles that are not sterile.

An alternative to high oral doses of iodine is to provide physiological doses much more frequently. The product usually used is called Lugol's iodine solution. One drop of Lugol's iodine, undiluted, contains about 6 mg of iodine. Lugol's iodine can be diluted so that perhaps 1 mg of iodine is consumed per person per week. If one drop of Lugol's iodine is put in 30 ml of water, then one teaspoonful of dilute solution will provide about 1 mg of iodine.

Preventing Iron Deficiency

Iron deficiency anaemia is the most prevalent of the three major micronutrient problems. It is the only one common in both industrialized and developing countries, and it is the most difficult of the three to control. For this reason the goal for the year 2000 is to reduce its prevalence markedly and not to eliminate it. This goal is achievable.

Iron nutrition is more complex than that of some other nutrients. Dietary iron comes in two main forms, namely haem and non-haem iron, which are not equally well absorbed and utilized; various dietary components adversely influence the absorption of iron from the intestine; and other substances such as vitamin C enhance the absorption of iron.

Unlike iodine and vitamin A deficiency, iron losses are caused by a highly prevalent parasitic infestation: hookworm disease. Some 800 million persons worldwide, mainly in developing countries, harbour hookworms and are therefore at risk of iron deficiency because the hookworms cause blood and iron loss. Schistosomiasis is another parasitic disease that causes loss of blood and therefore loss of iron in the urine or in the faeces. As in vitamin A deficiency, infections also contribute to iron deficiency, but they are not as prevalent or important as hookworm. Thus the treatment and control of hookworm infections and schistosomiasis constitute an important part of the strategy to combat iron deficiency in many tropical and subtropical countries. This tactic is considered below in the section on public health actions.

Improving Diets

To reduce the likelihood of iron deficiency, dietary diversity with a good balance of foods is particularly important. A small amount of food of animal origin such as meat, poultry or fish (especially the liver of these animals) is very helpful. While not essential, intake of animal products can greatly improve iron status. Cereals (such as rice, maize and wheat) and legumes provide most of the iron for most people worldwide; however, the iron is in the non-haem form, and absorption of the iron may be relatively poor. Diets promoted to control nutritional anaemias will include increased consumption of iron, but also foods rich in folate and especially vitamin C, which increases iron absorption.

Several of the foods being promoted to help control iron deficiency are the same as those recommended to improve vitamin A status; thus in many countries steps to increase dietary diversity might aim at the same time to improve both iron and vitamin A nutritional status. The promotion of green leafy vegetables and fruits will help both. Green leafy vegetables are relatively rich sources of iron and vitamin C and very rich sources of carotene, so an increase in their consumption will provide more iron, will enhance iron absorption because of the vitamin C and will increase the intake of vitamin A.

Another dietary measure is to reduce the intake at mealtimes of substances such as tannin in tea which decrease iron absorption or utilization.

The iron in breastmilk is very well absorbed, especially when compared with the iron in cows' milk or in products such as infant formula or milk powder made from cows' milk. Thus protection, support and promotion of breastfeeding is a strategy to prevent iron deficiency while the baby is exclusively breastfed, as well as to maintain iron status after the child is put on home foods while breastfeeding continues perhaps for 18 to 24 months. Breastfeeding also delays the return of menstruation, often by eight or more months. Menstruation is a cause of blood and iron loss for women. Thus breastfeeding may help to protect some mothers against iron deficiency when more iron is lost in menstruation than in breastmilk.

Public health actions

A broad range of public health measures and hospital practices contribute to reducing iron deficiency and other nutritional anaemias. The first area is obstetric practices. The traditional midwife often delivers the baby so that after birth it is below, not above, the mother. Also, in traditional practice the umbilical cord is not cut until it stops pulsating, or at least not immediately after delivery as is the practice of Western-trained doctors and obstetricians. When these two traditional practices are followed, considerably more blood enters the baby and red blood cell and haemoglobin levels are increased. They should be standard practices. Putting the baby to the breast in the first 30 minutes after delivery stimulates the uterus to contract, and this also reduces blood loss. Blood loss for the mother means iron loss, and many women go into delivery in an anaemic state.

Another public health measure of great importance in many countries is the control of hookworm infestations. Other parasites may also contribute to anaemia, and their control will reduce its prevalence. These include schistosomiasis, which is a cause of blood loss in urine if the infection is Schistosoma haematobium and in the faeces if the infection is Schistosoma mansoni or Schistosoma japonicum. Malaria also causes anaemia, mainly a haemolytic anaemia because the parasite destroys red blood cells.

Control of hookworm disease as a means of reducing anaemia has been a relatively neglected strategy until recently. Hookworm can be cured with a single dose of an anthelmintic drug such as albendazole, whereas curing anaemia may require 100 or more doses of iron using ferrous sulphate or some other compound. The delivery system is much simpler and problems of compliance do not exist. Deworming not only prevents chronic blood loss in the stools, but also improves the growth and appetite of children; if appetite improves, food intake including iron and vitamin C intakes may increase. In endemic areas treatment should be given at least once a year, while other public health measures are introduced to control transmission. These include health education and improved sanitation and water supplies.

The prevalence of nutritional anaemias is also influenced by the availability of birth spacing services. Pregnancy and childbirth increase iron needs and therefore contribute to anaemia. Some family planning methods that help prevent pregnancy such as abstinence, condoms or contraceptive pills thus contribute to control of iron deficiency. In contrast, intra-uterine devices (IUDs) in most women increase menstrual and other uterine blood losses and may thus contribute to anaemia.

Iron and folate supplementation, discussed below as a separate strategy, is usually considered a public health action. Nutrition and health education is also important in controlling iron deficiency.

Fortitication of foods with iron

Fortification of a wide variety of foods with iron has been feasible and used for many decades. In industrialized countries many different purchased foods are enriched with iron, especially cereal-based products. Unfortunately, fortification is much less used in developing countries where iron deficiency is particularly prevalent.

Fortification offers a very important strategy for control of iron deficiency in almost all countries North and South. If iron deficiency is to be substantially reduced before the year 2000, much more attention needs to be given to fortification, usually combined with the other strategies discussed here. Studies and various dietary surveys may be needed to determine the extent to which iron intake, iron bio-availability and other factors are the major causes of anaemia and to determine which foods are widely consumed and lend themselves to fortification. Several foods may be fortified with iron (whereas it is recommended that iodine should be added only to salt), but careful monitoring and quality assurance are essential.

Iron is not an easy nutrient to add to foods in a form that is well utilized and does not alter the quality of the food. The difficulty is to find a form of iron that is adequately absorbed and yet does not adversely influence the taste, colour or other attributes of the food being fortified. Unfortunately, ferrous sulphate, which is cheap and well absorbed, will often react with food constituents and cause colour changes. Iron phosphate does not have these negative attributes, but it is poorly absorbed. Sodium iron EDTA (ethylene-diamine-tetra-acetate) has recently been used successfully in Guatemala and elsewhere. It seems to lack the negative features of other preparations, and the iron is well absorbed. In Guatemala the vehicle for sodium iron EDTA has been sugar.

Many different foods have been fortified with iron and therefore offer possibilities for any country. These include wheat, wheat flour and bakery products, rice, maize flour, salt, sugar, condiments (such as fish sauce in Thailand) and processed foods. Chocolate-flavoured milk with added iron has been successfully used to control iron deficiency in children in Mexico.

Thirty years ago two research projects in Tanzania, one to investigate the causes of anaemia and the second to evaluate school feeding, used a powdered meat product manufactured in Kenya. This method was more or less abandoned until very recently, when animal haemoglobin was again suggested as a food additive or fortificant. Its advantage is that small amounts of haem iron will greatly increase the absorption of the good quantities of non-haem iron provided by a cereal-based diet.

Nutritionists and public health workers interested in the reduction of iron deficiency should advocate the fortification of foods with iron and perhaps also with vitamin C, folate and vitamin A.

Medicinal iron supplements

In many countries the main strategy for reducing iron deficiency is medicinal iron supplementation. The most common supplementation programmes provide or prescribe iron only for pregnant women. The coverage is sometimes extended to lactating women, but usually only at one postnatal visit soon after delivery. These programmes miss pregnant women who do not attend antenatal clinics, pregnant women before their first visit to the antenatal clinic, most breastfeeding mothers, females at risk prior to their first pregnancy and between pregnancies and all other iron-deficient people (or those at risk of iron deficiency) including children and adult males. Research in Kenya showed that 50 percent of primary school children and 40 percent of adult male road workers had low haemoglobin levels. Clearly iron deficiency is not confined to pregnant women.

Most iron supplementation programmes worldwide use ferrous sulphate, which is very cheap and provides iron in a form that is well absorbed. It is usually given in tablets providing 60 mg of elemental iron, and women are advised to take three tablets per day throughout pregnancy. Sometimes the use of this regime in antenatal clinics is allied with health and nutrition education to influence clinic attendance, in part to reduce anaemia. Ferrous sulphate is often combined with folate; such a product is often supplied to countries by UNICEF.

Problems have been encountered with compliance. It is reported that many women do not take the tablets because of perceived adverse reactions such as constipation, abdominal pain and black stools. Antenatal centres, clinics and health centres often run out of tablets, or health workers fail to give them out even if they are included in essential drug supplies.

There is a need to expand the use of iron supplements beyond pregnant women to include lactating women, females before pregnancy and between pregnancies, premature infants or those with low birth weight and, depending on the circumstances, certain preschool- and school-age children and some male adults.

Two important recent developments may change the way in which iron supplements are recommended. The first, and the less important, of these is the development and availability of slow-release iron capsules or spansules. These are made so that the iron, often ferrous sulphate, is slowly released in the intestine. Their advantages are that one rather than three doses is taken daily and some of the adverse reactions are reduced.

The second change is based on limited studies reported in 1993 which suggest that iron taken once per week is as effective as iron taken three times per day. Therefore it may soon be recommended that pregnant women and all others who can benefit from medicinal iron supplements be advised to take their iron supplement once per week, not three times per day. If one ferrous sulphate tablet providing 60 mg of elemental iron taken each week or every five days proves to be sufficient, iron supplementation will be easier, more acceptable to the public and much cheaper.

The control of malaria will also reduce anaemia, but it is not discussed here because it is not strictly a nutrition intervention. For many tropical countries malaria is the most important health problem, and it is a major cause of child deaths. Malaria does cause anaemia, but unlike the anaemia resulting from hookworm infection, it is not strictly a nutritional anaemia. In heavy infections, with massive parasitaemia, malaria parasites rupture millions of red blood cells. This causes haemolysis, the haemoglobin being released into the blood serum. Treatment of cases of malaria and the control of malaria transmission deserve a very high priority.

Simultaneous attention to several micronutrient deficiencies

There is great merit in combining actions to deal with several deficiencies at the same time. In particular, basic action to improve household access to and consumption of varied and adequate diets will help to control all micronutrient deficiencies. The fact that multiple benefits are achievable through food-based strategies is another reason why such interventions as home gardening, improved local food processing and nutrition education are the approaches of choice.

Family Feeding, Group Feeding and Street Foods

Most of the food that people consume in rural areas is eaten at home. This is also true in many urban areas, although street foods or foods eaten in stalls are providing an increasing percentage of food for urban dwellers. Inadequate family diets and family feeding are the fundamental causes of malnutrition in Africa, Asia, Latin America and elsewhere. For those who live away from home, particularly in institutions such as boarding schools, prisons or refugee camps, malnutrition or undernutrition may result from poor institutional diets.

In households, the two most important ways of procuring food are own food production, most commonly on small farms in rural areas, and purchase of food using money earned from work at home or outside the home or from sale of farm-produced products (termed cash crops, although they may be food crops such as cereals, fruits and vegetables put up for sale). Families or households may also procure food in other ways; these include take-home food donations, rations provided in exchange for work (food-for-work) and provision of supplementary foods to vulnerable groups. Foods may also be provided to households by friends or families as gifts or donations.

Feeding the Family

Food-insecure households are those where there is often an insufficient amount of food to satisfy the energy requirements, and also the energy wants or desirable allowances, of family members. There are other households, perhaps the majority, where for most of the year there is adequate food to assuage hunger, to fill everyone's stomach and at most times to satisfy energy wants. However, this "sufficient" food may comprise predominantly bulky, carbohydrate-rich foods and be very low in micronutrient-rich foods. As described elsewhere, bulk and insufficient frequency of feeding may result in energy intakes too low for requirements of young children, even if the food is available. In addition, poor appetites may reduce intake.

Different family members have different nutrient requirements which depend to some extent on age, gender, size, activity and other factors. Meals should provide adequate food to ensure that each family member receives all that is necessary to meet his or her nutrient requirements.

Cereals such as maize, rice, millet or wheat, if lightly milled, will usually provide both sufficient energy and B vitamins, although in the case of maize not enough to prevent pellagra. Foods other than the staple must provide the extra protein, fat, calcium, iron and vitamins A and C required. Africans, Asians and Latin Americans usually obtain adequate vitamin D from the action of sunlight on the skin. Iron may be almost sufficient in quantity from staple foods but is not in an easily utilizable form.

The extra protein required may come from protein-rich vegetable foods such as beans, groundnuts, cowpeas, soybeans, lentils or other legumes. Some may come from animal products such as meat, fish, milk and eggs. If the main foodstuff in the diet is cooking bananas, cassava, sweet potatoes or some other starchy food, then an even greater quantity of additional protein is necessary than in a diet based on a cereal grain.

A mixture of vegetable foods eaten at one meal, such as a cereal and a legume (e.g. maize or millet and cowpeas) or a root, a cereal and a legume (e.g. cassava, sorghum and groundnuts), provides better-quality protein than larger quantities of a single vegetable food would; the mixture usually contains all the essential amino acids, whereas a single cereal, root or legume is usually deficient in one or more of the essential amino acids.

A diet containing good quantities of legumes and occasional animal protein foods in addition to a cereal, banana or root staple probably satisfies the family's requirements for energy, iron, protein and B vitamins. It also provides fat if the legumes include good quantities of groundnuts or soybeans or if the animal protein consists of fatty meat, fish, milk or eggs.Such a diet is lacking only in vitamins A and C, which can best be supplied by fresh fruit and vegetables. Dark green leaves also provide much iron and some calcium.

Where foods of animal origin are not often available, the quality of the protein in the diet can be improved by providing a mixture or variety of vegetable products at each meal. Thus if a household has maize and beans available it is far better nutritionally to eat some of both at each meal rather than to eat maize for two weeks and then beans for two weeks. Mixtures of vegetable products are frequently eaten by many people in Africa and Latin America. Examples, of which some are traditional dishes, include:

— rice and bean stew;
— maize tortillas and beans;
— maize, beans and green leafy vegetables;
— baked sweet potatoes served with peas or beans and amaranth leaves;
— rice and lentil dhal;
— sorghum gruel and bananas with groundnut paste;

— millet served with onion, yam, peas and tomato relish;

— baked beans on wheat toast;

— plantains with beans and vegetables.

The various foods do not have to be physically mixed together, but can just as well be eaten separately at the same meal.

Family Feeding of Infants and Young Children

The importance of introducing foods to supplement breastfeeding when an infant is six months of age has been stressed. There are of course innumerable other recipes. For each family the foods used will depend on local customs, food preferences, food availability and cost.A wire sieve is useful in preparing infant foods. It serves to transform a solid or lumpy food into one of fine, soft consistency suitable for a child with few or no teeth. If a sieve is not available, one can easily be made by punching 20 to 30 holes in the bottom of an empty tin with a medium-sized nail. This makes a perfectly adequate sieve, but it should be carefully washed after each use.

Many adult dishes, after being put through a sieve, are suitable and good for young children. It must be remembered, however, that spices, especially those that have a burning hot taste, are unsuitable. Dishes containing curry powder, hot peppers, etc. should be avoided.

Breastfeeding should, under most circumstances, continue for as long as possible. An infant who has developed satisfactorily should begin supplementary feeding by the sixth month. A gruel of the local staple with added milk is an excellent food on which to start mixed feeding. If milk is not available, then any legume can be used instead. The supplementary food should at first be given in one feeding a day using a spoon and cup. After a week or two, when the child has become accustomed to semi-solid food, other dishes can be introduced. Next might come mashed fruit (e.g. mashed papaya) or vegetables, or possibly tomato or orange juice. A week or two later, some different foods, such as groundnut soup or bean mush. can be tried while the others are continued. At this stage, semisolid food might form part of two feedings a day.

By the end of the first year, all or any of the types of food in the recipes can have been tried, while breastfeeding continues. The infant should also by this time have been given the experience of trying many of the adult dishes of the family, with the exception only of obviously unsuitable foods such as peppery curries and alcoholic beverages.

During the period from 12 to 24 months, the infant can cope with many family dishes, but he or she should receive more frequent meals than adults and should have proportionately larger quantities of dietary fat, proteins and some other nutrients. A number of the suggested recipes can continue to supplement the family food and breastmilk that the toddler receives.

After the second year breastfeeding has usually ceased and extra energy- and nutrient-rich foods are important. The child is now able to cope with most of the family food but must get more than would appear to be the child's fair share. The young child may need more frequent feeding than the adult members of the family. Meals consisting mainly of starchy foods can be made more energy dense by the addition of a little oil or fat.

Institutional Feeding

There are many types of institutions where people receive food. The most important of these are schools, because at any one time many hundreds of millions of children worldwide are attending school. Most attend for part of the day at primary and secondary schools, where meals may or may not be provided. At boarding schools, where children sleep overnight, meals should be provided to supply all the nutrients needed for health and growth. In institutional feeding the main dish should be based on what is normally eaten in the country. This may be boiled rice, maize tortillas, ugali or wheat pasta.

In all of the institutional diets for which examples are given below, small additions of beverages or foods that are locally liked or traditionally eaten can be added. In many parts of the world such additions might include tea, coffee or other beverages. Another addition might be a particular relish, spice or other product to make the food more tasty or palatable, such as salsa, chutney, jam, honey or tomato sauce. Many dishes made in particular countries need or benefit from certain additional foods such as tomato paste, garlic, green peppers or relish.

Nursery Schools, Day-care Centres and Kindergartens

Many countries have an increasing number of nursery schools, day-care centres and kindergartens which are established as pre-primary schools or where children from one to six years of age can be left while their mothers work. Children attending such institutions should receive a daily meal consisting of food rich in those nutrients likely to be deficient in the home diet.

Table 1. Sample primary day school meals

Food	*Amount (g/person/meal)*
Example 1	
Make or rice	200
Mixed vegetables	50
Green leaves	25
Beans or groundnuts	100
Sugar	10
Milk (full-cream powder)	10
Oil (red palm)	10
Salt	5
Example 2	
Bananas (plantains) or potatoes	400
Groundnuts	50
Mixed vegetables	50
Beans	50
Dried skimmed milk powder (DSM)	20
Oil (red palm)	10
Salt	5
Example 3	
Cassava flour	150
Millet	100
Meat or fish	50
Beans	100
Green leaves	75
Fruit	100
Oil (red palm or other)	10
Salt	5
Example 4	
Bread	150
Potatoes	150
Tomatoes	75
Onions	50
Beans	100
Fruit	75
Oil	10
Salt	5

Older preschool children should be given a properly balanced midday meal similar to those suggested below for children in primary schools. Every effort should be made to provide nutrition education for the mothers who bring their children to these institutions. Mothers could be asked to help with meal preparations and would thus get first-hand experience in preparing nutritious dishes for young children. For a secondary day school the same foodstuffs could be provided, but the amount of each item should be increased by about 25 percent because older, taller and heavier children have increased nutrient requirements. It is not true that for a school lunch to be nutritious it must include a hot dish. Heat has nothing to do with nutritive quality. A cold lunch may be as nutritious as a hot lunch. It is the food served that determines the nutritive value of the meal.

School feeding and school food production can be linked very usefully with classroom activities relating to biology, health, home economics, geography, mathematics and agriculture. For example, practical lessons such as how to weigh and calculate amounts of foods in school meals and how to determine area and crop yields of school gardens can link mathematics to nutrition.

School feeding can be associated with school health services, which do not exist for many primary schools. It is useful to provide health examinations, some level of primary health care and first aid, to follow children's height and weight and to test their sight and hearing. A way should be found to make certain that children are immunized. There is also now beginning to be a movement to see that schoolchildren if necessary are regularly dewormed (where intestinal parasites are prevalent) and are provided with nutrient supplements such as iron, vitamin A and iodine.

There are many different reasons for and benefits of school feeding. These include preventing children from feeling hungry, which also helps them concentrate and benefit from classes; providing extra nutrients to improve children's nutritional status; improving attendance; and perhaps making it easier for mothers to work away from home, to be more productive in their fields or to increase their income.

Increasing data from research suggest that short-term hunger in schoolchildren who are not fed adequately before or at school adversely influences school performance, including learning and performance on psychological development tests.Schoolchildren, even quite young ones, can assist school feeding in a number of ways. They can help bring water to the feeding site if there is no running water at the school; carry wood or other fuel to the school; help in food preparation, in maintaining good food hygiene and in keeping the feeding area and utensils clean; and participate in school garden or small animal production activities.

Boarding Schools

Four examples of suitable boarding school diets are given in Table 2. The quantities of food given are suitable daily amounts for a secondary school pupil; these quantities should be divided up and served as three meals. Items such as meat, for which small daily amounts are indicated, may be given perhaps twice a week in larger amounts. For example, 20 g of meat per day make 140 g per week, which can be given in two equal portions of 70 g on Sunday and Wednesday.In a primary boarding school, the same foods could be provided but with an overall reduction in quantity of about 25 percent because younger children have lower requirements than older children.

Table 2.Sample secondary boarding school diets

Food	*Amount (g/person/meal)*
Example 1	
Make, rice or wheat (or mixture)	600
Beans	150
Groundnuts	100
Meat	20
Dried fish	20
Leafy vegetables	150
Fruit	100
Sugar	30
Dried skimmed milk (DSM) powder	20
Oil (red palm)	20
Salt	10
Example 2	
Bananas (plantains)	600
Potatoes	400
Rice	150
Meat	20
Beans	150
Groundnuts	50
Mixed vegetables	150

Fruit	100
Sugar	50
DSM powder	50
Oil (red palm)	10
Salt	10
Example 3	
Cassava flour	300
Millet	150
Green leaves	150
Fruit	100
Cowpeas	150
Groundnuts	100
Fish	50
DSM powder	50
Oil (red palm)	10
Salt	10
Example 4	
Rice	300
Potatoes	150
Maize	100
Bread	150
Meat	50
Eggs	30
Vegetables (mixed)	150
Margarine	50
Sugar	50
Fruit	150
Lentils	75
Groundnuts	50
Jam	30
Oil (red palm)	20
Milk (fresh)	0.5litre
Salt	10

Hospitals

Hospital patients usually spend most of their time in bed. Their needs for energy are therefore lower than those of active persons of the same sex, age and weight. However, some may have increased nutritional requirements.

These include patients who entered hospital undernourished; those who are pregnant or lactating or have recently had a baby; and those with diseases that require a special diet or extra nutrients. A generally suitable diet is shown in Table 3.

Table 3. Sample hospital diet

Food	*Amount (g/person/meal)*
Bread	100
Maize	200
Rice	150
Meat or fish	100
Vegetables	150
Fruit	150
Legumes	100
Milk (full-cream powder)	75
Oil	20
Salt	10

Agricultural estates and industrial enterprises

In some cases large numbers of agricultural and industrial workers spend six to ten hours working some distance from eating establishments. Where possible, a midday meal should be made available. The employer should decide, in consultation with the workers, whether the meal should be provided free, at subsidized prices or in a canteen where foods are sold more or less at cost. Free or subsidized meals will encourage as many workers as possible to partake. Canteen meals for workers may be expected to result in a higher output of work, a healthier, more contented labour force and reduced absenteeism. It is therefore often an economic advantage for an employer to provide such a meal.

Other institutions

A prison should provide a completely balanced diet suitable for persons doing heavy work. The diet should be cheap and simple. In some countries the scale of prison rations is laid down by law. In the United Republic of Tanzania each prisoner receives one 50-mg tablet of niacin per week in addition to the prescribed diet, to prevent the occurrence of pellagra. A suitable prison diet is shown in Table 4.

Table 4. Sample prison diet

Food	*Amount (g/person/meal)*
Maize rice, wheat or millet	750
Beans	150
Vegetables	150
Groundnuts	100
Meat	20
Sweet potatoes	50
Fruit	100
Salt	10
Oil	5

A diet that might be served in the army is shown in Table 5.

Table 5. Sample army diet

Food	*Amount (g/person/meal)*
Make, rice or wheat products	400
Bread	100
Potatoes	400
Beans	100
Vegetables (mixed)	150
Onions	25
Groundnuts	100
Fruit (fresh)	200
Fruit (dry)	50
Meat	250
Milk (fresh)	0.5 litre
Sugar	60
Oil	50
Salt	10

Vulnerable Group Feeding

The term "vulnerable" is better used, if it is applied to those at special risk of malnutrition. Thus vulnerable children may include those who do not have evidence of malnutrition but are at risk for any of numerous different reasons; for example, they may include children from very poor families, children from large families with narrow birth spacing and in some cultures female children from a low caste. Similarly, it might be better to say that women of child-bearing age, rather than just pregnant and lactating women, are at risk, and then again to find criteria such as poverty, female-headed households and other factors that place them at risk. Other vulnerable groups include older people if not cared for by an extended family, individuals with mental illness and orphans who are not cared for by relatives. Some chronic diseases such as tuberculosis and acquired immunodeficiency syndrome (AIDS) make subjects very vulnerable to malnutrition.

Supplementary feeding of young at-risk children may be done at feeding centres to which mothers take the children or by providing take-home food supplements or even rations for a complete diet. Research on supplementary feeding has shown that often some of the food taken home is consumed by others and not all by the target child. If the household is food insecure, however, this food may help all members. Consideration might be given to providing for more than just the nutritional needs of the child.

Supplementary feeding has been used in growth monitoring programmes where the food is provided only to children under five years of age who have evidence of malnutrition. This approach has been used in Tamil Nadu, India. When the children show a defined degree of growth improvement, they become ineligible for further feeding. The approach is claimed to be very successful in rehabilitating children. However, feeding only malnourished children, rather than children at risk, constitutes a curative approach, whereas a preventive one would in general be preferable.

Supplementary feeding of children is much more likely to have an impact on nutritional status of populations if combined with sustainable agricultural development and efforts to reduce poverty, allied with primary health care, immunizations, education about treatment for diarrhoea and deworming.

Supplementary feeding has possible disadvantages which need to be appreciated. Families who receive food for their children may become overly dependent on free food and may not make adequate efforts to improve home food security. If rations are used as an incentive to motivate attendance at growth monitoring centres, then if supplements are not available, attendance might decline.

The same general principles apply to supplementary feeding of women or any other group. Some programmes provide dietary supplements to all pregnant and lactating women. Foods likely to reduce anaemia are of particular importance. Medicinal supplements of iron, or iron and folate, are often also provided. If women are to be selected on the basis of risk, then the first criterion should be the level of poverty. Other risk factors include teenage pregnancy; death or malnutrition of a previous child; chronic disease such as tuberculosis or AIDS; low weight for height in small women; and poor social support, especially in female-headed households.

Supplementary feeding should not be done in isolation. Primary health care needs to be offered and must be easily accessible; nutrition and health education should be provided; and consideration needs to be given as to whether to refer mothers to family planning services.

Food-for-work and take-home food rations

Food-for-work, where food allocations are made in return for work rather than as free donations, is often used in food emergencies such as drought or famine. Food-forwork can also be used in development programmes and other situations.

Increasingly food-for-work has been used as full or partial payment of wages, often for work done on public works programmes planned by a government; as an incentive for voluntary labour; and sometimes as a budgetary support for a developing country. Food-for-work is a strategy often used by the World Food Programme (WFP).

The kinds of work undertaken have usually involved labour-intensive road building projects; environmental projects, including tree planting and forestry; and projects to open up new land. Usually these projects are carried out in areas with food shortages. They have both nutritional and non-nutritional objectives: to help prevent food insecurity, hunger and malnutrition, and to get good public work done. The internationally supported food-for-work programmes are usually also seen as economic assistance to developing countries.

Where food is given in exchange for work, it is highly desirable to provide some level of primary health care for workers and to give advice on nutrition, on how to prepare foods not normally eaten by those receiving the food and on what other foods besides the donated rations would help balance the diet.

Each programme establishes the amounts and kinds of foods to be given. These decisions should be based on sensible criteria and concern for local food habits. Often 2100 kcal are provided per person, but food needs to be provided to satisfy family needs, not just the workers' needs. Often five daily rations of 2100 kcal per day worked are provided as a family ration.

A typical ration would provide 400 to 500 g of cereal flour or rice, 25 to 50 g of legumes or pulses and 25 to 35 g of oil or fat. If available, about 20 g of meat or fish might be added. Fresh perishable foods are not usually included in the rations, and it is important that workers and their families supplement the ration with fruits and vegetables. There is some debate about the energy level; a higher level than 2100 kcal may sometimes be advocated.

Street Foods

Although the term "street foods" has only recently become widely used, the sale and consumption of food on city streets goes back many centuries. Now street foods are recognized as having a very large role in urban food consumption, especially in developing countries and for the poor and middle classes. FAO studies have shown that in some countries street foods provide a very significant proportion of total food intake for many people. It is surprising that the nutritional, health, social and economic impact of street foods has not been studied or appreciated until relatively recently. FAO has had a leading role in drawing attention to the importance of street foods; the Organization has held conferences on the topic and provided advice to countries on appropriate measures to make these foods safer for the consumer. Because of its expertise in this area, FAO can provide very useful advice and assistance to member countries. A good deal of the following discussion has drawn on FAO publications and papers relating to street foods.

FAO has defined street foods as follows: "Street foods are ready-to-eat foods and beverages prepared and/or sold by vendors especially in streets and other similar public places". This definition is now widely accepted. Street foods are mainly sold in urban areas, but they are also prepared and sold by vendors under similar circumstances in rural areas, and not strictly on the street. It has become increasingly common for entrepreneurs to set up simple facilities or stalls adjacent to rural schools or to work under a nearby tree to prepare and sell ready-to-eat foods and drinks to schoolchildren and other passers-by. These foods have the same advantages and risks as urban street foods.

In developing countries the street food phenomenon has greatly mushroomed in recent years, in parallel with the huge increase of people living in urban areas, including the vast and ever-expanding megalopoli in Asia and Latin America and rapidly expanding cities everywhere. Street foods are also sold extensively in industrialized countries; it is not unusual for the New York banker or the London journalist to purchase a hot dog and a soft drink or a bag of fish and chips, respectively, on the street.

In cities in developing countries street foods provide a significant percentage of the total food intake of millions of people, have an important economic role and employ many persons, yet these activities are largely unregulated and create risks to health.

Even though authorities in many countries, North and South, regard food vendors on the street as generally undesirable and a cause of problems for the cities, the fact is that street vending has a vital role: urban workers and dwellers depend on it, it is a major employer, it contributes to the city economy and it is a major source of food for many people. Food vendors have also become an important part of urban social life and are not infrequently an attraction to the city.

City officials concerned about problems or potential problems caused by street food vendors should seek to resolve the problems, not drive the vendors off the streets. There are ways to improve the safety of street foods. Authorities should recognize that street foods are generally popular because they provide an accessible source of relatively cheap food of a kind desired by busy urban people such as factory and office workers, students, shoppers and travellers. Especially in the middle of the day, very few people can return home to eat. Street foods are also convenience foods: they save the homemaker or single person from cooking and perhaps from fuel gathering. Many poor people in crowded housing do not have proper cooking facilities, so the food vendor may provide breakfast, lunch and dinner. An FAO study revealed that in Bangkok street foods contributed 88 percent of the daily energy, protein, fat and iron intake of children aged four to six years.

In most countries the street food industry, even though very extensive and important economically, is considered part of the informal economy. It usually does not get much official or positive recognition. Consequently, governments and cities have not taken the necessary steps to improve the quality and safety of foods sold or generally to regulate the practice. The street food industry requires recognition, as it is often very large, involves large amounts of money, employs huge numbers of people and provides a real service to many citizens. Regulations should come at the same time as recognition. The industry is one of the few that can be entered with very little capital, relatively little education and only a small amount of expertise. Success requires hard work, ingenuity and street wisdom. These are characteristics of many unemployed people and of some who enter illegal parts of the informal economy. In many countries such as Nigeria and Indonesia, the majority of persons employed as street vendors are women, so the sector contributes to empowerment and economic gains for females.

Before the year 2000, there is surely a need for governments to recognize that street food vending is not a temporary phenomenon that will be replaced when development is successful. It may have undesirable features, but there are many positive aspects for cities and nations. What is now needed is recognition, legalization and improvement.

The objectives of regulation and control are to improve the quality and safety of foods consumed and to let the industry have a positive role in city life. The difficulty is the risk that overregulation could drive the industry underground, force up prices and cause loss of jobs. Sensible steps must recognize the service provided by the industry. A prescription cannot be provided for the regulation and control of street foods in all countries. Appropriate regulatory activities must take cognizance of national differences, culture, local law and current street food practices.

In countries that do not have any regulations, the first step might be to recognize publicly the existence of street food vendors and to issue statements on the importance of the industry and its problems. The second step might be to map out and count the vendors and to classify them using some locally appropriate system. The third step might be to provide each vendor with an official licence. Usually a fee would have to be paid to obtain the licence. The fee should not be so high as to drive vendors away or underground, but it could go towards funds to assist in upgrading the hygiene and other practices of vendors.

It is necessary also to decide on certain minimum standards to help reduce health hazards. These standards will depend on local circumstances and probably should be established after consideration by a committee on which vendors and consumer associations have some representation. Regulations need to be appropriate and to fit in with national and city policies and legislation, and they must be aimed at improving the wholesomeness and safety of food sold without greatly raising the prices. They must have no negative impact of any significant degree either on employment or on the economy, and they must not greatly reduce the availability of street foods enjoyed by the public.

In some countries the first regulations introduced have been unnecessarily stringent health requirements for vendors which have contributed little to protection of the public. Regulations that should be considered might address the cleanliness of the facility, the quality and quantity of water used and the training of vendors regarding appropriate food handling practices in order to reduce the risk of contamination.

Regulations are only effective if there is some monitoring system and some surveillance. Inspections are useful and should be used not entirely for punitive purposes, but also for educational opportunities. Trained inspectors should be able to record violations and possibly to threaten or take action, but they should also be used to make constructive suggestions for improvement of vendor practices or for upgrading the stall or facility. Guidelines for the design of control measures for street-vended foods in Africa are being developed by the Codex Alimentarius Commission along these lines.

An advantage of recognition, licensing and regulations is that they move the street food industry out of the strictly informal sector and into the formal sector. This may make it possible for vendors to get credit or loans to improve their operations. A disadvantage, especially if licence fees are high and regulations strict, is that many vendors will attempt to evade licensing while continuing to practise their trade. There is also a strong possibility that inspectors will be bribed to close their eyes both to non-licensed vendors and to contraventions of the regulations.

Singapore, a highly regulated country with a strong economic base, chose to resettle its food vendors into particular market areas or centres and to issue licences dependent on health standards. Singapore street foods are undoubtedly cleaner than those elsewhere in Asia, but they are perhaps less convenient for the public, less amenable as social meeting places and more likely to be run by persons not in the lower strata of society.

Food hygiene and sanitation

It does not take knowledge of tropical medicine or epidemiology to appreciate that foods prepared and offered for sale on the streets of many cities in developing countries present a health hazard. Anyone who appreciates that germs cause disease can realize that food that has been touched by dirty hands and utensils, that is not served very hot and that has been covered with flies may be unsafe.

Foods may be contaminated not only with pathogenic organisms such as viruses, bacteria and parasites which cause human disease, but also by dangerous levels of food additives, toxins, residues of pesticides used in food production or preservation or other poisonous substances such as heavy metals, e.g. lead, which is toxic.

There have been reports of deaths or disease from consumption of street foods from many countries; instances have included 14 deaths in Malaysia from consuming rice noodles, 300 people becoming ill in Hong Kong from consuming food that apparently contained a toxic pesticide, and a cholera outbreak in India.

Contamination results from unhygienic practices in the preparation, cooking, serving and storage of food. Food vendors, unlike well-run restaurants, often lack refrigeration, good storage facilities and efficient stoves. Bacteria may be in the food when it is purchased, but they are likely to multiply if the food is not refrigerated or properly stored. Organisms in food may be destroyed by the heat of cooking, but if the food is not thoroughly heated and well cooked they may infect the person who eats the food. Some organisms produce toxins in food. The problems are usually related either to lack of coldness or refrigeration for food storage or lack of heat to cook the food. The other risk factors contributing to food contamination are lack of cleanliness of the premises, the utensils and the food handlers. After preparation and cooking, foods may be contaminated by unwashed hands; by flies, cockroaches, rodents and dust; and by holding at temperatures that encourage explosive bacterial reproduction.

A major problem for street vendors which then produces a health risk for the client is water. Ice can also be a source of infections; vendors may use ice made from contaminated water. In some countries it is rare for street food vendors to have running water, and it is common for vendors to have to fetch water from a considerable distance from their point of operation. Then there is a temptation to use the water sparingly, because getting water takes time and energy. The water carried to the stall or other facility may be clean and safe, or it may be contaminated. Some food sellers on the street may have water that is not potable, and they very frequently have an inadequate supply of water. Water is essential for cooking many foods, for washing foods, for making some beverages or for drinking, for cleaning cooking vessels and utensils and for vendors to wash their hands. Vendors may not have hot water for washing utensils. Not infrequently an operator will rinse off utensils for hours in one bucket of water which becomes increasingly dirty and contaminated. All of these practices greatly enhance the likelihood that organisms such as salmonella, shigella and Escherichia coli will be transmitted and that certain parasitic infections such as giardia and ascariasis will be spread.

The food hygiene problem is made worse by the fact that most vendors have very little knowledge or appreciation of the importance of safe, hygienic food handling. City authorities may not take steps to control the unhygienic practices of food vendors because the officers who have authority on the streets may not themselves be aware of the risks. Many of the consumers of street foods also have little knowledge of or interest in food hygiene. This compounds the problem, because they may not, for example, insist on foods being well heated or select food stalls that appear cleaner.

Another health risk to the public is that street food vendors are often unable to dispose properly of waste water and refuse. Often there is no good system for disposal of garbage, which may end up in the street or gutter. Similarly, used water may not go into a drain but may accumulate around the stall or in puddles on the street, where it may attract flies and mosquitoes which may breed and spread disease. Singapore has strict regulations requiring vendors to use plastic garbage bags and provides metal bins near to where hawkers sell food.

Precautions that an informed purchaser can take are to insist that the food be very hot, that it come from the heating area to the customer very quickly and that it be served on a clean plate. Food taken straight off the grill is likely to be safe. It is wise to select fruits that can be peeled just before eating, such as bananas, and to order a beverage from a bottle that can be opened just prior to drinking.

Nutritional quality of street foods

Relatively little research has been published on the nutritive value of street foods sold in different countries. If large numbers of people get 50 to 80 percent of their nutrients from street foods, then it is important that the foods be nutritious and provide a good proportion of the essential nutrients. The nutritional quality of street foods obviously varies enormously from country to country, but also from vendor to vendor in a large marketplace. On city streets around the world it is possible to choose a meal that is nutritious and well balanced, as well as very tasty.

People naturally select foods or dishes for purchase more on the basis of preference, price and affordability than on the nutrient content of the meal or on nutritional quality criteria. Clearly anything that can be done to improve the nutritional quality of foods sold on the street will be helpful.

References

Economic Commission for Africa (ECA)/FAO.1978. *Manual on child development, family life and nutrition,* by J.A.S. Ritchie. Addis Ababa, Ethiopia.

Engle, P.1992. *Care and child nutrition.* Paper for the International Conference on Nutrition. New York, USA, United Nations Children's Fund (UNICEF).

FAO.1976. *The feeding of workers in developing countries.* FAO Food and Nutrition Paper No. 6. Rome.

Latham, M.C. & van Veen, M.1989. *Dietary guidelines.* Proceedings of an international conference, Toronto, Canada, 1988. Cornell International Nutrition Monograph Series No. 21. Ithaca, New York, USA, Cornell University.

United States National Academy of Sciences, Food and Nutrition Board.1989. *Recommended dietary allowances.* Washington, DC, USA.

WHO. 1989. *Dietary management of young children with acute diarrhoea. A practical manual for district programme managers,* by D.B. Jelliffe & E.F.P. Jelliffe. Geneva, Switzerland.

Bibliography

Aguilera, Jose Miguel and David W. Stanley. 1999. *Microstructural Principles of Food Processing and Engineering*. Springer.

Alleyne, G.A.O., Hay, R.W., Picou, D.I., Stanfield, J.P. & Whitehead, R.G. 1977. *Protein-energy malnutrition.* London, UK, Arnold.

Berg, A.1987. *Malnutrition. What can be done? Lessons from the World Bank experience.* Baltimore, Maryland, USA, Johns Hopkins University Press.

Brown, M.L. 1990. *Present knowledge in nutrition.* Washington, DC, USA, International Life Sciences Institute, Nutrition Foundation. 6th ed.

Brun, T.A. & Latham, M.C. 1990. *Maldevelopment and malnutrition.* World Food Issues, Vol. 2. Ithaca, New York, USA, Cornell University, Program in International Agriculture.

Cannon, G.C.1992. *Food and health: the experts agree.* London, UK, Consumers' Association.

Carpenter, Ruth Ann; Finley, Carrie E. 2005.*Healthy Eating Every Day*. Human Kinetics.

Dawson, R.J. & Canet, C. 1991. International activities in street foods. *Food Control,* 2(3): 135-139.

Economic Commission for Africa (ECA)/FAO.1978. *Manual on child development, family life and nutrition,* by J.A.S. Ritchie. Addis Ababa, Ethiopia.

Engle, P.1992. *Care and child nutrition.* Paper for the International Conference on Nutrition. New York, USA, United Nations Children's Fund (UNICEF).

FAO. 1992. *Integrating diet quality and food safety into food security programmes,* by M.F. Zeitlin & L.V. Brown. Nutrition Consultants' Reports Series No. 91. Rome.

FAO/WHO. 1992b. *International Conference on Nutrition. Nutrition and development a global assessment.* Rome.

Gibson, R.S. 1990. *Principles of nutritional assessment.* Oxford, UK, Oxford University Press.

Gopalan, C. & Kaur, H.1993. *Towards better nutrition - problems and policies.* Special Publication Series No. 9. New Delhi, India, Nutrition Foundation of India.

Hetzel, B.S.1989. *The story of iodine deficiency: an international challenge in nutrition.* New York, USA and Oxford, UK, Oxford University Press.

Holland, B., Unwin, I.D. & Buss, D.H.1988. *Cereals and cereal products. Third supplement to McCance & Widdowson's The composition of foods.*Nottingham, UK, Royal Society of Chemistry.

James, W.P.T. & Schofield, E.C.1990. *Human energy requirements: a manual for planners and nutritionists.* Oxford, UK, Oxford University Press/FAO.

Katz, Solomon. 2003. *The Encyclopedia of Food and Culture,* Scribner.

King, F.S. & Burgess, A.1993. *Nutrition for developing countries.* Oxford, UK, Oxford University Press. 2nd ed.

Latham, M.C. & van Veen, M.1989. *Dietary guidelines.* Proceedings of an international conference, Toronto, Canada, 1988. Cornell International Nutrition Monograph Series No. 21. Ithaca, New York, USA, Cornell University.

Marion Nestle, 2007. *Food Politics: How the Food Industry Influences Nutrition and Health,* University Presses of California.

Maxwell, S. & Frankenberger, T.R. 1992. *Household food security: concepts, indicators, measurements. A technical review.* New York, USA, United Nations Children's Fund (UNICEF)/International Fund for Agricultural Development (IFAD).

McGee, Harold. 2004. *On Food and Cooking: The Science and Lore of the Kitchen.* New York: Simon and Schuster.

McLaren, D.S. 1983. *Nutrition in the community.* New York, USA, John Wiley and Sons. 2nd ed.

Passmore, R. & Eastwood, M.A. 1986. *Davidson and Passmore human nutrition and dietetics.* Edinburgh, Scotland, UK, Churchill Livingstone. 8th ed.

Sanjur, D.1982. *Social and cultural perspectives in nutrition.* Englewood Cliffs, New Jersey, USA, Prentice-Hall.

Shils, M.E., Olson, J.A. & Shike, M. 1994. *Modern nutrition in health and disease.* Philadelphia, Pennsylvania, USA, Lea and Febiger. 8th ed.

Souci, S.W., Fachmann, W. & Kraut, H. 1989. *Food composition and nutrition tables 1989/90.* Stuttgart, Germany, Wissenschaftliche, Verlagsgesellschaft.

UN ACC/SCN. 1991b. *Managing successful nutrition programmes.* Nutrition Policy Discussion Paper No.8. Geneva, Switzerland.

United States National Academy of Sciences, Food and Nutrition Board.1989. *Recommended dietary allowances.* Washington, DC, USA.

Waterlow, J.C. 1992. *Protein energy malnutrition.* London, UK, Edward Arnold.

WHO.1976. *Methodology of nutritional surveillance.* Report of a Joint FAO/UNICEF/ WHO Expert Committee. WHO Technical Report Series No. 593. Geneva, Switzerland.

World Bank.1994. *Enriching lives. Overcoming vitamin and mineral malnutrition in developing countries.* Washington, DC, USA.

World Health Organization (WHO). 1966. *The assessment of the nutritional status of the community, by* D.B. Jelliffe. WHO Monograph Series No. 53. Geneva, Switzerland.